Walter Edward Collinge

The Skull of the Dog

A Manual for Students

Walter Edward Collinge

The Skull of the Dog
A Manual for Students

ISBN/EAN: 9783741187049

Manufactured in Europe, USA, Canada, Australia, Japa

Cover: Foto ©Klaus-Uwe Gerhardt /pixelio.de

Manufactured and distributed by brebook publishing software
(www.brebook.com)

Walter Edward Collinge

The Skull of the Dog

THE

SKULL OF THE DOG.

A MANUAL FOR STUDENTS.

WITH A GLOSSARY OF OSTEOLOGICAL TERMS.

BY

WALTER E. COLLINGE, F.Z.S.,

ASSISTANT LECTURER AND DEMONSTRATOR IN ZOOLOGY AND COMPARATIVE
ANATOMY, MASON COLLEGE, BIRMINGHAM.

Illustrated.

London:

DULAU & CO., 37, SOHO SQUARE.

Birmingham:

CORNISH BROTHERS, 37, NEW STREET.

1896.

BIRMINGHAM:

PRINTED BY ROBERT BIRBECK & SONS, BROAD STREET.

PREFACE.

THE present Manual has been written with the view of supplying the student of Comparative Anatomy and Osteology, with a concise and accurate account of the structure of a Mammalian skull. For this purpose, that of a dog has been chosen, it being now largely used for general teaching purposes ; and it has also the advantage of being easily procurable in both the adult and young stages.

A short glossary of osteological terms has been appended, which it is hoped will add to the usefulness of the work.

The illustrations are all from original drawings, made by Mr. F. W. Crispe and myself; and drawn on wood and engraved by Mr. Edwin Wilson of Cambridge, to whom my best thanks are due for the great pains he has bestowed upon the same.

I take the opportunity of expressing my thanks to Professor T. W. Bridge for several valuable suggestions, which have materially increased the value of the work ; to Professor B. C. A. Windle, who has very kindly permitted me to reprint the tables concerning the cranial characters of the different breeds of dogs, in Chapter IV ; and to Mr. T. Manners-Smith, and Mr. W. A. Brockington for assistance in revising the proof sheets.

W. E. C.

Mason College,
Birmingham, 1895.

ERRATA.

Page 5, line 5 from bottom, for "cell sometimes" *read* "cell is sometimes."

,, 24, in foot-note 1, for " periotic anchylose" *read* "periotic bones, &c."

,, 25, in foot-note 1, for " Beever " *read* " Beaver."

,, 36, line 4, for "cornu" *read* "cornua."

,, 44, line 7 from bottom, for "cup" *read* " cap."

,, 51, line 6 from bottom, for " maxilla, pre-maxilla," *read* " maxillæ, pre-maxillæ."

,, 58, for " Weidersheim " *read* " Wiedersheim."

CONTENTS.

LIST OF ILLUSTRATIONS.

CHAPTER I.

INTRODUCTION.

The Study of Osteology is one, which the student of Comparative Anatomy is apt to neglect; and it is only quite recently that it has been adequately represented in the ordinary course of Zoology and Comparative Anatomy, taught in the University and College Classes of this country.

It is of the greatest importance, that a definite and accurate knowledge should be obtained of the framework, supporting and modifying the tissues which cover it. Further, as has been pointed out by Professor Flower, "large numbers of animals, all of those not at present existing on the earth, can be known to us by little else than the form of their bones."

The student, who, having mastered the facts laid down in the following pages, desires to acquire a further knowledge of general osteology, cannot do better than carefully study the admirable work by Professor Flower and Dr. Gadow, and Parker and Bettany's "Morphology of the Skull." References to a number of memoirs upon the cranial osteology of

B

various mammals will be found cited in the references to Chapters III and IV, which may be consulted with advantage by senior students.

Preparation of Specimens. — To obtain perfect skeletons or parts of skeletons, the larger portion of flesh, &c., should be carefully dissected away. The part should then be placed in water to macerate for a time ; after which any remaining portions of the flesh may be brushed away by a small brush with stiff bristles : they should then be well washed and bleached by exposure to sunlight. During the process of bleaching, which will take two or three weeks, the bones should be repeatedly rinsed with water.

When sections of a bone or series of bones, such as the skull, are required, a fine saw should be used. Young skulls it is best to imbed in a matrix of plaster of Paris, as the sutures have not anchylosed, and the parts are apt to fall away. In making longitudinal sections, the skull should be cut slightly to the left of the median line, so as not to damage the mesethmoid or vomer. If microscopic sections are required, either of the following methods should be adopted :

1.—The method I prefer for decalcification is to soak the bones in a solution of nitric acid and alcohol, three parts of nitric to seventy parts of alcohol ; in this solution they are allowed to remain for several days, or weeks, according to their size and age. Sections should be stained in a weak aqueous solution of eosin for about five minutes, dehydrated in absolute alcohol, and mounted in Canada balsam (benzol-solution). Young or fœtal bones may be treated as follows : (Busch's method). Place in a mixture of bichromate of potash 1 per cent, and 1-10 per cent of chromic acid ; decalcify in 1 or 2 per cent solution of nitric acid, to which a small quantity of chromic acid (1-10 per cent), or chromate of potash (1 per cent), has been added.

2.— Sections of non-decalcified bone may be obtained by the following method (Ranvier). When the flesh has been removed, the bone should be cut into lengths, and allowed to macerate in water, which is being constantly changed. When all the soft parts are destroyed, the bone should be dried ; and the sections cut with a fine saw. The section should then be taken, and rubbed down on moistened pumice-stone. Both sides of the section should be rubbed smooth and then polished on a moistened Turkey hone. Spongy bone should be soaked in gum or copal solution, and allowed to dry, before rubbing down.

CHAPTER II

THE STRUCTURE AND FORMATION OF BONE.

Connective Tissues.—Those tissues, whose function it is to serve as connecting, supporting, or skeletal substances are known as **connective tissues**; and include such forms as areolar, elastic, fibrous, retiform, lymphoid and adipose tissue, cartilage and bone; all of which are derived from the mesoblast. It is with these two last mentioned substances—cartilage and bone—that we are at present more directly concerned.

Cartilage, in mass, is a yellowish or bluish-white elastic tissue, the cells of which are imbedded in a firm ground substance, or matrix. Excepting on those surfaces articulating with some other part of the body, it is usually covered by a fibrous connective tissue the **perichondrium,** which is richly supplied with blood-vessels, lymphatics and nerves. No nerves pass into the substance of the cartilage; it is, therefore, non-sensitive.

Cartilage contains a substance known as *chondrigen,* which yields on boiling *chondrin,* a mixture of mucin, gelatin, and various salts in small proportions. Chondrin is soluble in hot water, and insoluble in alcohol, chloroform, or ether.

According to the character of the matrix or ground substance all cartilage may be grouped under three varieties: 1. **Hyaline cartilage**, 2. **Elastic cartilage**, and 3. **White fibro-cartilage.**

1. **Hyaline cartilage** forms the costal cartilages of the sternal portion of the ribs, the skeleton of the embryo, &c.

It takes its name from the hyaline transparent nature of its ground substance.

The cells, which are irregularly scattered throughout the ground substance, are oval or spherical nucleated masses of protoplasm, enclosed in a transparent membrane or capsule, which forms what is known as the **cartilage lacunæ**. This membrane is not usually distinguishable from the ground substance, except when treated with special reagents. In each lacuna there is a single cell ; but two, four, or more may be present. In certain forms of hyaline cartilage, fine channels connect the lacunæ with one another, and enter into connection with the lymphatics of the perichondrium.

The protoplasm of the cartilage cell is sometimes filled with globules of fat.

It has been shown by Tillmans, Babor, and others that, under special treatment, hyaline cartilage can be broken up into fibres.

2. **Elastic Cartilage** known also as yellow, reticular, or spongy cartilage, is opaque and more flexible than either hyaline or fibro-cartilage. The ground substance is traversed by a dense branched network of elastic fibres, with small spaces around the cells. Between the meshes of these fibres, and in the space surrounding the cells, hyaline cartilage is present. Elastic cartilage occurs in the epiglottis, Eustachian tube, larynx, and ear.

White-fibro cartilage is much more flexible, and tougher than hyaline cartilage. It consists of series of bundles of fibres arranged in layers, in which there are irregularly distributed cartilage cells. Where the fibres are very dense, these cells become very flattened. White fibro-cartilage occurs in the intervertebral and interarticular discs, sesamoid cartilages, &c.

Development of Cartilage.—Like the rest of the skeletal tissues of the body, cartilage arises from the mesoblast. In the embryo, certain cells of the mesoblast assume a definite shape, usually polygonal ; and a transparent substance surrounds each cell, which ultimately forms the capsule. In this stage there is no matrix. Later, the cells enlarge and divide, and an intermediate matrix is formed. On the division of a cartilage cell, a new capsule is formed around the new cells ; the old capsule blending with, and ultimately forming part of the matrix. The process of division is as follows : the original cartilage cell divides into two, a new capsule being formed around these, the old one blending with the matrix ; these two again divide, now forming a group of four, each having a separate capsule ; they may further divide, and form a group of eight. The origin of the capsule is but imperfectly known ; whether or not it is secreted by the cartilage cell (Kölliker), or formed from a part of the protoplasm of the cell, is not yet settled.

Bone is a connective tissue, the ground substance of which has become impregnated with various calcium salts ; and which has entered into association with other tissues to form a support for the softer parts of the body.

Bone may be either *compact*, or *spongy* or *cancellous*, the two imperceptibly passing into each other. Compact bone is of an ivory nature, and forms the peripheral portion of the bone. Spongy or cancellous bone forms the articular portions of long shaft-like bones. In the bones of the skull—tabular bones— the spongy tissue lies between an inner and outer layer of compact tissue. In such bones the cancellated layer is termed *diploë*.

The whole of the skeleton is invested by a fibro-vascular membrane, the **periosteum.** It consists of two layers : an outer-fibrous layer, a series of closely aggregated bundles of white fibres ramified by blood-vessels; and an inner or **osteogenetic layer** consisting of a looser network of elastic fibres. Beneath this second layer is a fibrous network containing, in the young bone, numerous protoplasmic nucleated cells of a spherical or oblong form ; these are the **osteoblasts** —the active agents concerned in the formation of new bone. Those portions of the bones, which enter into articulation with others, are covered by hyaline cartilage ; the periosteal layer being absent here.

The external portion of the bone, the matrix or osseous substance, is a dense fibrous connective tissue. The matrix—**ossein**—is largely impregnated with various insoluble inorganic calcium salts, chiefly phosphates and carbonates.

The osseous element is arranged in a series of microscopic plates, or **lamellæ,** between which are a series of oblong or oval spaces, the **lacunæ.** From each of these spaces numerous fine canals—**canaliculi** —pass out, and fuse with those passing out from the lacunæ above and below. The lacunæ and canaliculi together form the lymph canalicular system, and communicate with the lymphatic vessels of the marrow.

In each lacuna there is a flattened protoplasmic cell, containing an oval nucleus termed the **bone cell.** In the young bone the cell gives off pseudopodia-like branches, which pass into the canaliculi. In the adult condition these branches are few ; and the cell is then often termed a bone corpuscle.

In an ordinary shaft of bone (compact bone) the

lamellæ are arranged either in a concentric manner around a blood-vessel **(Haversian canal)**, or as interstitial or ground lamellæ.

Each Haversian canal contains two blood-vessels, an artery and vein, surrounded by connective tissue. The canals pass in the bone in a longitudinal direction, giving off transverse or oblique branches ; and so anastomose with one another. They open on to the osteogenetic layer of the periosteum by fine pores, and into the medullary cavity ; as the canals approach the medullary cavity, they gradually enlarge, and finally fuse with the marrow tissue. The concentric lamellæ surround the Haversian canals, while the interstitial or ground lamellæ fill in the space between the Haversian systems.

In the spongy portion of the bone, the cavities are termed Haversian spaces, and are usually filled with a red marrow ; the bone substance here forming a series of septa or lamellæ, which are termed the **bone trabeculæ.**

The centre or hollow cavity of a bone is termed the **medullary cavity;** surrounding this internal cavity in the form of a lining membrane is a vascular layer of areolar tissue the **endosteum.**

The marrow is a fibro-vascular tissue, consisting of a matrix of retiform and adipose tissue, blood-vessels and cells. The marrow cells are termed yellow marrow and red marrow respectively. The former fills the cavities of tubular bones ; the latter occurs at the ends of tubular bones, and in the spongy bone substance. There are also present in the marrow a series of various sized multi-nucleated cells, the **myeloplaxes** of Robin ; which, according to Kölliker, are concerned with the process of bone absorption, while other authors regard them as entering into the formation of blood corpuscles.

Formation and Growth of Bone.—Bone is developed in two ways, viz., in the cartilage of the embryo, or directly from the osteogenetic layer of the periosteum ; these two methods are known respectively as **intracartilaginous** or **endochondronal** and **intramembranous** or **periosteal.**

The future skeleton of the embryo arises as a series of mesoblastic cells ; these undergoing certain modifications, and becoming cartilaginous, and later impregnated with calcareous matter. It has been pointed out, however, that bone arises by another mode, distinct from the skeleton of the embryo, and at a later period. It will, therefore, be well, to first consider that first formed in the embryo.

1. **Intracartilaginous** or **endochondronal ossification.** The ossification of a cartilage, such for instance as that which in the adult will form the humerus, commences in the middle, the ossification gradually proceeding towards the ends.

The cells in this central portion of cartilage first

become enlarged, and then separated some little distance from one another, by the surrounding matrix. Calcareous matter is deposited in this matrix which later encloses the cartilage cells. In the meantime the cartilage cells have become much larger and flatter, and assumed a definite arrangement, forming a series of columns radiating from the centre. While this central ossification has been taking place, a growth of osseous substance has been formed on the surface of the cartilage, immediately beneath the periosteum. It has already been pointed out (p. 8) that, beneath the inner layer of the periosteum, there are a series of nucleated cells termed osteoblasts ; and it is due to these that this osseous layer on the outer surface of the cartilage is formed. As layer upon layer is formed, lacunæ containing osteoblasts are also formed. After a time this outer or osteoblastic tissue becomes irrupted, and passes into the inner ossified portion ; ultimately filling up the areolæ, and finally entirely supplanting the ossified cartilage which is gradually absorbed.

2.—**Intermembranous** or **endochondral ossification.** Covering the embryonic cranium, we find a tissue or membrane composed of a series of fibres—osteogenic fibres—large granular corpuscles and an intervening ground substance. Calcium carbonate is deposited within these fibres, which form bony spicules ; these extend in all directions, in certain regions becoming thicker by the deposition of bony matter, and form a small bony plate. By continued growth, this plate extends, until it meets with neighbouring bones. What, in the adult skull, appear as sutures are, in the embryo, filled by a vascular connective tissue, in which numerous osteoblasts are present. The growth of this tissue ceases, when the bones have become complete, and only the sutures remain ; and even these in old skulls may become almost obliterated.

SPECIAL REFERENCES.

PIZZOZERO : " Neue Untersuchungen ueber d. Bau des Knochenmarkes bei den Vögeln."
Arch. f. mikr. Anat., 1890, T. xxxv.

SCHÄFER, E. A.: " Quain's Elements of Anatomy."
Vol. 1, part ii, " General Anatomy or Histology." London, 1891.

STRICHT, O. v. d. : " Recherches sur la structure fondamentale du tissu osseaux."
Arch. de Biologié, 1889, T. ix.

STRICHT, O. v. d.: "Recherches sur la cartilage hyalin."
Arch. de Biologié, 1887, T. vii.

THIN, G.: " On the Structure of Hyaline Cartilage."
Proc. Royal Soc., 1885.

CHAPTER III

THE DOG'S SKULL.

The Mammalian Skull may conveniently be divided into a cranial and facial portion : the former consisting of the brain-case, or cranium proper, and the auditory bulla ; the facial portion being composed of those parts in front of or below the orbits, including the jaws and olfactory capsules.

THE CRANIUM.

The bones of the cranium proper are grouped in three segments or rings, viz., (i) the **occipital,** (ii) the **parietal,** and (iii) the **frontal** segments.

  a. The **occipital segment** consists of three cartilage bones, which in old animals may fuse together inseparably.

    i. The **basi-occipital** is a median flat bone, bounded anteriorly by the basi-sphenoid, laterally by the right and left auditory bulla, and posteriorly by the inferior margin of the foramen magnum.

    ii. The **ex-occipitals** form the lateral boundaries of the foramen magnum, and the occipital condyles. In front of the condyles, and close to the posterior lateral borders of the basi-occipital, they are perforated by the **condylar foramina,** through which the hypoglossal nerve passes. The ex-occipitals are each laterally produced into a prominent downward and outwardly directed process, the **par-occipital** or **paramastoid process,** which enters into close relationship with the auditory bulla, and serves for the attachment of the digastric and other muscles. Between the ex-occipital and periotic, the **foramen lacerum posterius** passes, through which the ninth, tenth, and eleventh nerves find their exit.

iii. The **supra-occipital** is a large, median, and somewhat convex-shaped bone, forming the superior portion of the occipital segment ; anteriorly it passes into the parietal segment as a narrow prominent ridge, to which the splenius and other muscles are attached; the squamosal also forms the lateral portion of this ridge, known as the **lambdoid** or **occipital crest,** to which part of the temporal muscle is attached.

b. The **parietal segment** consists of both cartilage and membrane bones. Its superior and inferior margins meet posteriorly with the occipital segment; laterally, they are separated by the interposition of the squamosal, and the auditory bulla.

i. The **basi-sphenoid** is a median flat triangular bone, the base of the triangle being posterior and lying immediately in front of the basi-occipital. Laterally, it is bounded by the pterygoids, alisphenoids, and squamosals. On the dorsal surface of the basi-sphenoid is a little hollow, the **sella turcica,** which lodges the pituitary body.

ii. The **ali-sphenoids** arise from the sides of the basi-sphenoid as two wing-like bones, forming part of the floor of the cranial cavity. From the ventral surface of each ali-sphenoid, there projects a downwardly directed vertical plate of bone, the **pterygoid process,** whose anterior edge is connected with the palatine. Between the posterior portion of the ali-sphenoids, and the ex-occipital, the **foramen lacerum medium** passes; and through it the internal carotid artery enters the cranial cavity. The basal portions of the ali-sphenoids are perforated by the **foramen ovale,** a large oval-shaped foramen transmitting the third division of the trigeminal nerve (vth), and, anterior to this, the **foramen rotundum,** a much smaller and more circular aperture, through which the second division of the trigeminal nerve passes. Between the ali-sphenoids and orbito-sphenoids, is a large oval foramen the **foramen lacerum anterius,** or **sphenoidal fissure;** through it the motor-occuli (iii) pathetic (iv) and abducens (vi) nerves find exit.

iii. The *inter-parietal* is a small narrow median bone, anchylosed with the supra-occipital, and extending for some distance between the parietals.

iv. The *parietals* form the greater portion of the upper and lateral surface of the cranium; they articulate with each other in the mid-dorsal line, forming the **sagittal suture;** the point of suture forms a raised crest, the **sagittal crest,** which divides in the anterior portion of the parietals and extends anteriorly on either side of the frontals, as far as the post-orbital process of these bones; it serves for the attachment of the temporal muscle, which passes through the space bounded by the zygoma, known as the **temporal**

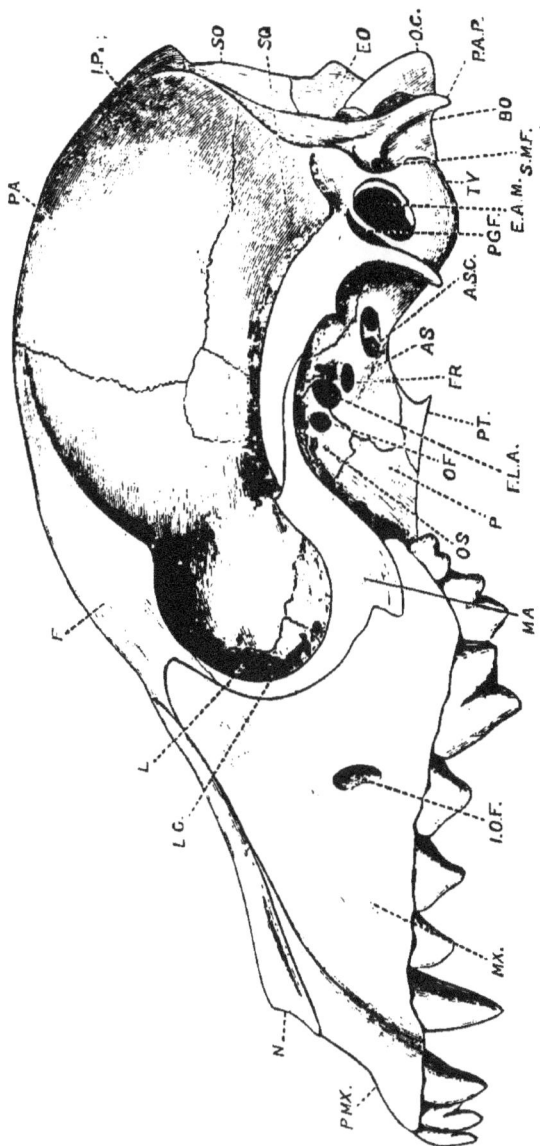

Fig. 1.—*Canis familiaris*. **Lateral View of the Skull.** (F.W.C.)

A S. Ali-sphenoid. **AS.C.** Ali-sphenoid canal, posterior opening. **B O.** Basi-occipital. **E.A.M.** External Auditory meatus. **E O.** Ex-occipital. **F.** Frontal. **F.L.A.** Foramen lacerum anterius. **F.R.** Foramen rotundum. **I O.F.** Infra-orbital foramen. **I P.** Inter-parietal. **L.** Lachrymal. **L C.** Lachrymal canal. **M A.** Malar. **M X.** Maxilla. **N.** Nasal. **O.C.** Occipital condyle. **O F.** Optic foramen. **O S.** Orbito-sphenoid. **P.** Palatine. **P A.** Parietal. **P G.F.** Post-glenoid foramen. **P.M X.** Pre-maxilla. **PA.P.** Par-occipital process. **P T.** Pterygoid. **S M.F.** Stylomastoid foramen. **S O.** Supra-occipital. **S Q.** Squamosal. **T Y.** Tympanic bulla.

C

fossa, and converges in a fan-like manner to its point of insertion into the **coronoid process** of the mandible.

c. The **frontal segment** forms the anterior boundary of the brain case ; it lies anterior to the parietal segment, with which it articulates.

> i. The **pre-sphenoid** is a median bone in front of the basi-sphenoid, bounded laterally by the pterygoids, and the posterior portion of the palatines. Its ventral or basal portion is straight, with a somewhat irregular dorsal portion. It is cancellous or spongy in texture. Its posterior and dorsal portion forms the lower margin of the **optic foramina,** through which the optic nerves pass out ; anterior to this it diverges on either side of the median line, and forms part of the inner walls of the orbits.

> ii. The **orbito-sphenoids** are a pair of bones, forming part of the wall of the orbits and cranium. They are fused with the pre-sphenoid, and united suturally with the frontals, ali-sphenoids, and squamosals. Their posterior borders form the upper margin of the **optic foramina.**

> iii. The *frontals* are a pair of membrane

bones, forming the roof and sides of
the anterior portion of the cranium,
meeting both dorsally and ventrally.
They unite suturally; posteriorly with
the parietals by the transverse **coronal
suture,** ventrally with the orbito-sphenoids and pre-sphenoids, and anteriorly with the jugals, lachrymals, maxillæ, and nasals. The most anterior
portion is produced, and passes between the nasals and maxillæ. Laterally each is slightly produced downwards and outwards into a process
above the orbit, the **supra-orbital
process.**

iv. The *lachrymals* are two small bones
forming part of the anterior walls of
the orbit. They lie between the frontals, maxillæ and jugals. Each is
perforated by a small aperture for the
passage of the **lachrymal duct.** The
lachrymal is probably homologous
with the antorbital bone of fishes,
which is the most anterior of a series
of sensory canal bones bordering the
orbits.

d. The **ethmoidal region** is situated anterior to
the cranium, and forms the dorsal and posterior division of the facial portion of the skull.

i. The **mes-ethmoid** consists of two parts,

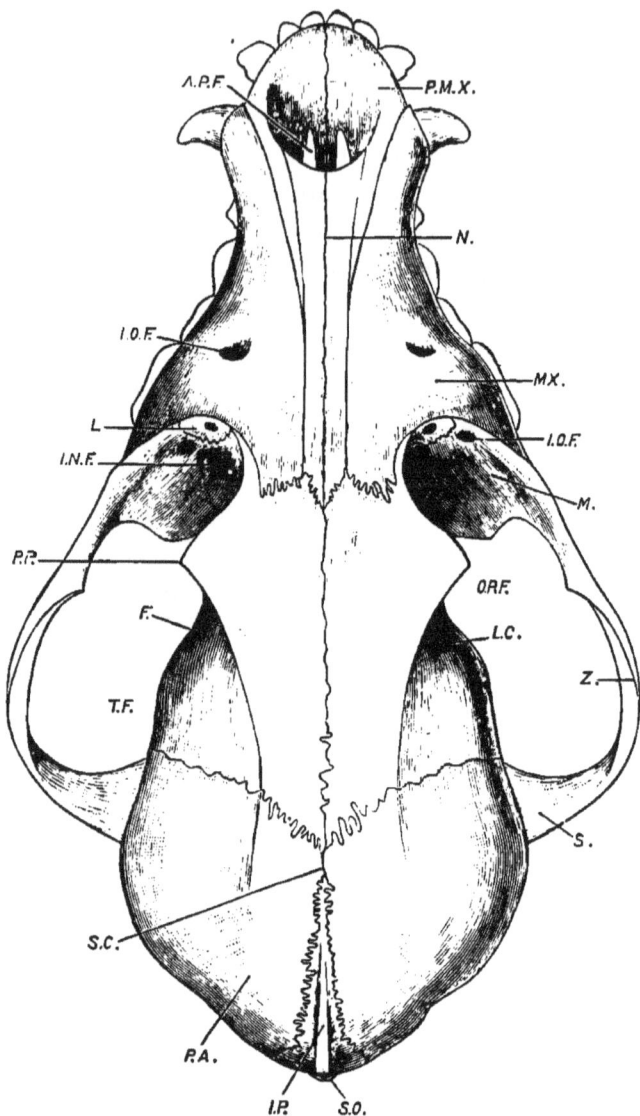

Fig. 2.—*Canis familiaris.* **Dorsal surface of the Skull.** (F.W.C.)

A.P.F. Anterior palatine foramen. **F,** Frontal. **I N.F.** Inter-orbital foramen. **I O.F.** Infra-orbital foramen. **I P.** Inter-parietal. **L.** Lachyrmal. **L C.** Lambdoidal crest. **M.** Malar. **M X.** Maxilla. **N.** Nasal. **O R.F.** Orbital fossa. **PA.** Parietal. **P. M X.** Pre-maxilla. **P P.** Post-orbital process. **S.** Squamosal. **S C.** Sagittal crest. **S O.** Supra-occipital. **T F,** Temporal fossa. **Z.** Zygomatic arch.

an upper and posterior portion, which
· forms the inferio-anterior portion of
the brain-case, the **cribriform plate.**
It is perforated with numerous small
holes, through which the divisions of
the olfactory nerve pass to the nose.
The lower and anterior portion, the
lamina perpendicularis forms a vertical plate immediately in front of the
cribriform plate: its anterior border
is produced and forms a vertical
cartilaginous plate, the **septum nasi ;**
the two forming a partition between
the right and left olfactory cavities.

THE SENSE-CAPSULES.

At an early stage in the development of the dog,
the olfactory, optic, and auditory organs become
enclosed from their ventral surface by cartilage;
the lateral walls of the basal plate extend upwards and inwards, ultimately enclosing the
whole of the sense organs, and, in some vertebrates, part or all of the dorsal surface. Such
a cartilaginous skull, however, does not persist
in any of the mammalia, but in some fishes,
e.g. Elasmobranches, it forms the skull (chondocranium) of the adult. In the higher Vertebrata, cartilaginous capsules ossify from certain
centres and form bones. The position, form,

&c.,of these ossifications in the dog are described below.

a. The **olfactory capsule** in the dog forms a very large portion of the facial region, entering into very close relationship with the ethmoidal region, and the bones of the maxillary arch.

   i. The *nasals* are two long flat membrane bones, forming the dorsal portion of the nasal cavities. Posteriorly they articulate with the frontals, and with each other in the median line.

   ii. The **naso-turbinals** form the inferior surface of the nasals; they are thin laminated bones, and are produced into pouch-like processes.

   iii. The **ethmo-turbinals** consist of a series of infolded laminæ of bone, lying in the posterior and upper portion of the nasal cavity. They are fused with the lower portion of the cribriform plate.

   iv. The **maxillo - turbinals** are similar bones to the ethmo-turbinals; they occupy the anterior portion of the nasal cavities. They are thinner, and altogether much more delicate, than the ethmo-turbinals, and folded much closer.

   v. The **mesethmoid cartilage** lies upon

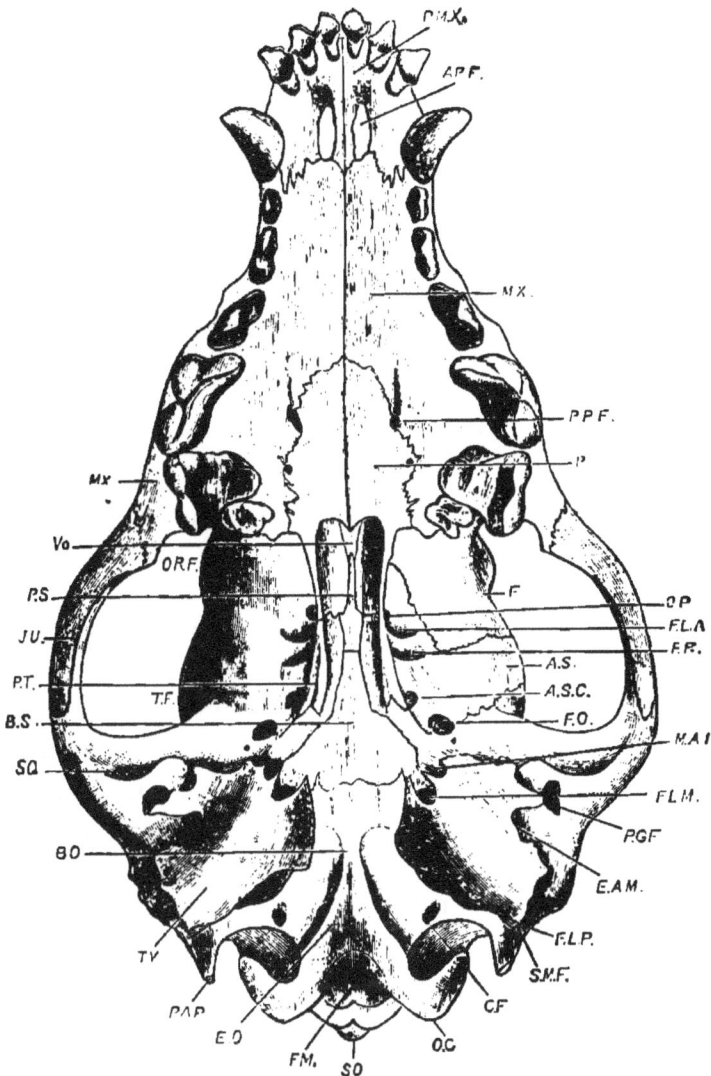

Fig. 3.—*Canis familiaris.* **Ventral Surface of the Skull.** (F.W.C.)

A.P.F. Anterior palatine foramen. **A S.** Ali-sphenoid. **A S.C.** Ali-sphenoid canal, posterior opening. **B O.** Basi-occipital. **B S.** Basi-sphenoid. **C F.** Condylar foramen. **E.A.M.** External auditory meatus. **E O.** Ex-occipital. **F.** Frontal. **F.L.A.** Foramen lacerum anterius. **F.L.M.** Foramen lacerum medium. **F.L.P.** Foramen lacerum posterius. **F M.** Foramen magnum. **F O.** Foramen ovale. **F R.** Foramen rotundum. **M A.** Malar. **M.A.I.** Meatus auditorius internus. **M X.** Maxilla. **O C.** Occipital condyle. **O P.** Optic foramen. **O R.F.** Orbital fossa. **P.** Palatine. **P A.P.** Par-occipital process. **P G.F.** Post-glenoid foramen. **P S.** Presphenoid. **P T.** Pterygoid. **P.MX.** Pre-maxilla. **S M.F.** Stylomastoid foramen. **S O.** Supra-occipital. **S Q.** Squamosal. **T F.** Temporal fossa. **T Y.** Tympanic. **V O.** Vomer.

the dorsal border of the vomer, and forms part of the partition between the two nasal cavities ; in old skulls it is very largely ossified.

vi. The *vomer* is a long narrow blade-like bone, posteriorly bounded by the anterior portion of the pre-sphenoid ; its ventral border lies upon the maxilla and palatine. It terminates anteriorly a little behind the suture of the premaxillæ.

b. The **optic capsules.** Those bones forming the boundaries of the orbit have already been described in their respective segments.

c. The **auditory capsules.** In the embryo, as has been pointed out, the auditory organs are enclosed within a cartilaginous capsule which later becomes ossified, and is replaced by the otic bones. In the human skull ossification proceeds from three centres termed respectively **pro-otic, epi-otic,** and **opisthotic.** [1]

i. The **periotics** are the series of fused otic bones, already referred to, situated between the occipital bones and the

---

[1] "In birds, the three periotic anchylose with one another, as well as with the adjacent supra-occipital and exoccipital, so completly, that even the Y-shaped suture becomes obliterated." (Huxley).

In the crocodile the epiotic is early united with the supra-occipital, and the opisthotic with the ex-occipital ; the pro-otic alone remains separable throughout life." * * "In the turtle, the opisthotic remains permanently distinct." (Miall).

squamosal. They are divided into two portions : a **petrous portion,** arising from the pro- and opisthotic ; and a **mastoid portion,** arising from the epiotic.

The **petrous portion** forms the anterior part, and encloses the internal ear. On its outer surface are two foramina situated almost above each other : the larger is the **fenestra ovalis,** and the smaller the **fenestra rotunda;** together they form the means of communication between the internal ear, and the tympanic cavity. On the inner surface of the petrous portion of the periotic, there are two large fossa: the upper and larger one is the **floccular fossa** [1], which lodges the floccular lobe of the cerebellum ; the lower one has a ridge of bone passing across it, which divides it into an anterior and a posterior opening. The posterior one is the **meatus auditorius internus,** which transmits the auditory nerve to the internal ear. The anterior opening forms the

[1] Its size varies in different skulls. In many mammals it is entirely absent (*e.g.* Sheep, Elephant, *Echidna* &c.) In some of the Rodentia (Porcupine, Capybara) it is also absent, while in others (Beever, *Chinchella*) it is very conspicuous.

aqueductus Fallopii; through it the facial nerve passes.

The **mastoid portion** forms the posterior and external portion of the periotic. Its most posterior portion is directed downwards, and forms the **mastoid process;** between the mastoid and the auditory bulla is the **stylomastoid foramen,** transmitting the main trunk of the facial nerve.

ii. The *tympanics* lie on the outer side of the periotics, they are somewhat flask-shaped. The portion, which forms in nearly all the Carnivora such a conspicuous prominence on the ventral surface of the skull, is the **bulla:**[1] laterally it is slightly produced; its edges being uneven, and supporting the cartilage of the external ear. The lateral opening of the bulla, the **meatus auditorius externus,** has at its base an incomplete bony ring, the **tympanic ring,** which rises up as a part of the septum; in life it is covered by the **tympanic membrane.** The septum lies behind the tympanic ring, aud divides the bulla into an inner

---

[1] In *Thylicinus* (a Marsupial) whose skull resembles in very many particulars that of the Dog, the tympanics are rudimentary and do not anchylose with the cranial elements, and there is no bulla.

and outer chamber. From the outer chamber the **Eustachian canal passes.** Its connection with the ear, it should be remembered, is quite a secondary one, it forming no essential portion of the auditory organ. It arises as a modification of the embryonic hyomandibular cleft, persisting in certain fishes as a spiracle.

iii. The **auditory ossicles.** Morphologically, the chain of three small bones lying across the tympanic cavity from the tympanic membrane to the membrane across the fenestra ovalis, should be treated of together with the bones of the mandibular and hyoidean arches. Their very close relationship, however, to the auditory organ makes it more convenient that they should be examined here.

α. The **malleus** is a small stout bone, consisting of a body and two processes. The larger process, the **manubrium,** is very distinct, being long, thin, and strongly curved, and attached to the tympanic membrane. The smaller process articulates with the tympanic bone in a small fossa.

β. The **incus** articulates with the malleus, this portion being produced and form-

ing a saddle-shaped head. Posteriorly
a process extends on to the periotic
bone, in a fossa of which it articulates.
The incus is pyriform in shape; the
two processes forming the upper portion of the bone; the lower portion or
stalk bends downwards and inwards
and becomes attached to a minute
disc-like bone, the **os orbiculare.**

γ. The **stapes** is stirrup-shaped : ventrally
its base is attached to the membrane
covering the fenestra ovalis ; while its
upper portion, which forms the arch,
articulates with the os orbiculare.

With regard to the development and homologies of
the auditory ossicles, much uncertainty prevails.
The generally accepted views are as follows.
The malleus and incus arise in connection with
the mandibular arch, the former from the lower
portion and the latter from the upper portion
of the arch ; the malleus being homologous
with the articular and angular elements of
Meckel's cartilage, and the incus with the palato-
quadrate cartilage of lower Vertebrata. The
stapes arises in connection with the hyoid arch,
and an ossification from the periotic capsule ;
according to Kitchen Parker, Hertwig and
others, it is the homologue of the columella in
Birds, Reptiles, and Amphibia.

THE JAWS.

The maxillary and mandibular arches arise as modifications of the primitive cartilaginous visceral arches of the embryo ; thus becoming adapted to a function very different from their primitive one. In the downward growth of the first visceral arch, an upward and forwardly directed process is developed known as the maxillary process (palato-quadrate cartilage), which forms the foundation of the upper jaw of the adult ; growing forwards they meet with a cartilage, the naso-frontal process. The downwardly directed portions of the arch has in the meantime continued its growth, and become united by its distal end with the process of the opposite side, the two forming the future lower jaw (Meckel's cartilage).

a. The **maxillary arch.** The bones of the maxillary arch enter into very close relationship with those of the olfactory capsule, and the ethmoidal region of the cranium.

    i. The *pterygoids* are two plate-like bones lying upon the basi-sphenoid and ali-sphenoid. Their lateral borders are partly supported by the ali-sphenoids and palatines. Anteriorly they join with the posterior portions of the palatines. The posterior portion of

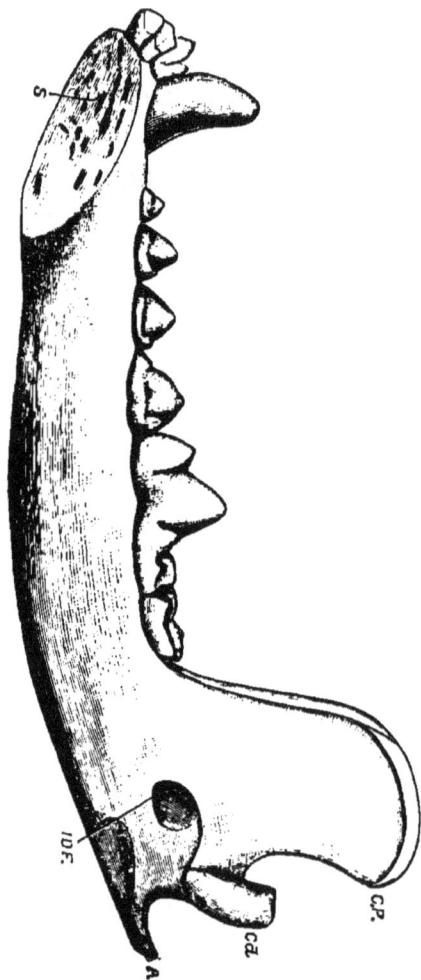

Fig. 4.—Canis familiaris. Inner side of right ramus of
the lower jaw. (F. W. C.)

A. Angle. Cd. Condyle. C P. Coronoid process. I.D.F. Inferior dental
foramen. S. Symphysis.

each is produced backwardly into a curved lamellar process, the **hamular process.** The median space between the two pterygoid plates is known as the **meso-pterygoid fossa.** Lateral to the hamular process is a small space, the **pterygoid fossa.**[1]

ii. The *palatines* are two bones lying anterior to the pterygoids and articulating with the maxillæ, the vomer, the orbital portions of the frontals, the lachrymals, the pre-sphenoid, orbitosphenoids, and ali-sphenoids. Each palatine bone consists of two almost vertical plates, termed respectively the **ascending** and **horizontal plates.** The former constitutes the main portion of the bone, and is perforated by the **posterior palatine foramen.**

iii. The *maxillæ* form the greater part of the facial portion of the skull, and consist of two large irregular shaped bones. Posteriorly they articulate with the jugals and frontals; anteriorly, with the premaxillæ. They carry the canine, premolar, and molar teeth. Above the first premolar is a large aperture, the **infra-orbital foramen,**

---

[1] More distinctly seen in the Cat's skull.

through which the maxillary division
of the fifth nerve passes.

Anteriorly each maxilla is produced into a
plate-like portion, the **facial** or **nasal
plate.** In a like manner posteriorly
each is produced and forms the **malar
process.** Behind and beneath the
malar process is a small **tuberosity,**
in which are a number of small foramina, for the passage of the superior
dental nerve and blood vessels. The
ventral portion is termed the **alveolar
portion ;** it is hollowed out into the
alveoli, or sockets, for the lodgment
of teeth. From the alveolar border
a large horizontal process passes inwards, the **palatine plate,** which meets
with its fellow of the opposite side in
the median line, in a **sutural ridge**
which lends support to the vomer.

iv. The *premaxillæ* form the most anterior
portion of the skull ; together with
the maxillæ, with which they articulate, they form the upper jaw. They anchylose with each other in the median
line, and also carry the incisor teeth.
The lateral and dorsal portions of each
bone articulates with the nasal.

On the ventral surface each is perforated
by a large aperture, the **anterior**

palatine foramen. Their median portion is produced into a folded laminæ, which lodges an accessory organ of smell known as **Jacobson's organ.**

v. The *jugals* or *malars* form the boundaries of the orbit below and in front. Dorsally they articulate with the frontals and maxillæ. Posteriorly each bone is produced into a laterally compressed bar, forming part of the **zygomatic arch.** They support the zygomatic process of the squamosal on their free ends.

b. The **mandibular arch.**

ı. The *squamosals* are two large bones wedged in between the occipital and parietal segments. They form the hinder portion of the sides of the cranial cavity, and articulate with the parietals and frontals.

From the ventral portion of each bone a strong outwardly directed process passes forwards, the **zygomatic process,** and meets with the posterior portion of the jugal; the two forming the **zygomatic arch.**

On the under surface of the squamosal, and on the inner border of the base of the zygomatic process, is an oblong surface, the **glenoid fossa** : the posterior edge is projected forwards and forms the **postglenoid process** ; the two serving for the articulation of the condyle of the mandible.

ii. The *mandible* consists of two rami joined anteriorly by a symphysis, and, posteriorly, articulating with the glenoid fossa of the squamosal bone. Each ramus consists of a short anterior portion, the **horizontal ramus,** lodging the teeth, and a thinner hinder angulated portion, the **ascending ramus.** On its upper border is a large **condyle,** which articulates with the glenoid fossa of the squamosal. Beneath this condyle is a vertically extended process, the **angle.** The anterior border bears a deep groove, its outer portion or **coronoid process** being turned inwards, and is the point of insertion of the temporal muscle.

On the inner side of each ascending ramus is a large foramen, which admits a branch of the mandibular nerve, the dental nerve, to the teeth. It lies just behind the last molar, slightly above and in front of the angular process ; and is known as the **inferior dental foramen.** It leads into the **dental canal.**

On the outer side of each horizontal ramus,
beneath the first premolar, is a smaller
opening, the **mental foramen,** through
which a branch of the dental nerve
makes its exit.

The point of junction of the two horizontal
rami is termed the **symphysis**; on the
lower border of which the digastric
muscle is inserted in a small depression.

The actions of the muscles moving the
lower jaw may be briefly indicated as
follows :

The lower jaw is raised by the masseter,
internal pterygoid, and temporal muscles ; the depression of the jaw is
accomplished by the digastric muscle,
and a series of hyoid muscles. The
larger portion of the external pterygoid muscle, having their fibres nearly
horizontal, draw the jaw forward ; while,
when these muscles act alternately, the
jaw is moved from side to side : this
movement, however, in the dog is
very restricted. The retraction of
the jaw is effected by the hinder part
of the temporal muscle.

THE HYOID BONES.

The hyoid bones serve for the attachment of the
root of the tongue and the larynx ; in the dog
they consist of a body, and two anterior and
posterior cornu.

    a. The **body of the hyoid.**

        i. The **basihyal** is a flat bar-like bone,
            having its terminal portions turned
            dorsally and thickened. It forms the
            ventral portion of the hyoidean
            apparatus.

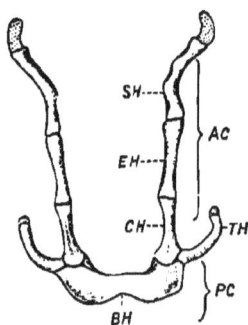

Fig. 5.—*Canis familiaris.* **Hyoidean apparatus, front view.**
(W. F. C.)

S H. Stylohyal. E H. Epihyal. C H. Ceratohyal. B H. Basihyal. T H.
Thyrohyal. A C. Anterior cornu. P C. Posterior cornu.

    b. The **posterior cornu.**

        ii. The **thyrohyal** articulates with the
            basihyal and is united with the
            thyroid cartilage of the larynx ; it

consists of a curved bone between these two points.

c. The **anterior cornu.**

Each consists of four parts, and connects the series with the cranium.

i. The **tympanohyal** lies in a canal on the inner side of the stylomastoid foramen between the tympanic and periotic bone. Its ossified portion is cylindrical and is connected with the proximal end of the **stylohyal** by a band of cartilage.[1]

ii. The **stylohyal** is a curved, somewhat twisted bone, forming the posterior portion of the anterior cornu, of which it is the largest division.

iii. The **epihyal** is slightly shorter than the stylohyal, and dilated at its distal and proximal ends.

iv. The **ceratohyal** is a short stout bone, attached at its distal end to the basihyal.

---

[1] As has been pointed out by Prof. Flower, "it can be seen more distinctly in some dogs' skulls than others."

FORAMINA AND APERTURES OF THE SKULL.

The foramina through which the cranial nerves make their exit from the cranial cavity, and through which certain blood vessels pass, are as follows:

i. The **foramen magnum,** a large median opening on the posterior end of the skull, in the occipital segment. Through it the spinal cord passes to the brain.

ii. The **condylar foramina** pass obliquely through the ex-occipitals ; at either side of the basi-occipital and anterior to the condyles, they transmit the **hypoglossal nerve.**

iii. The **foramen lacerum posterius** is situated in front of the condylar, and between the ex-occipital bone and auditory bulla ; through it the **glosso-pharyngeal, pneumogastric,** and **spinal accessory nerves** make their exit, the **internal jugular vein** also passes through.

iv. The **stylomastoid foramen** is a small irregular opening between the auditory bulla and the paroccipital process ;

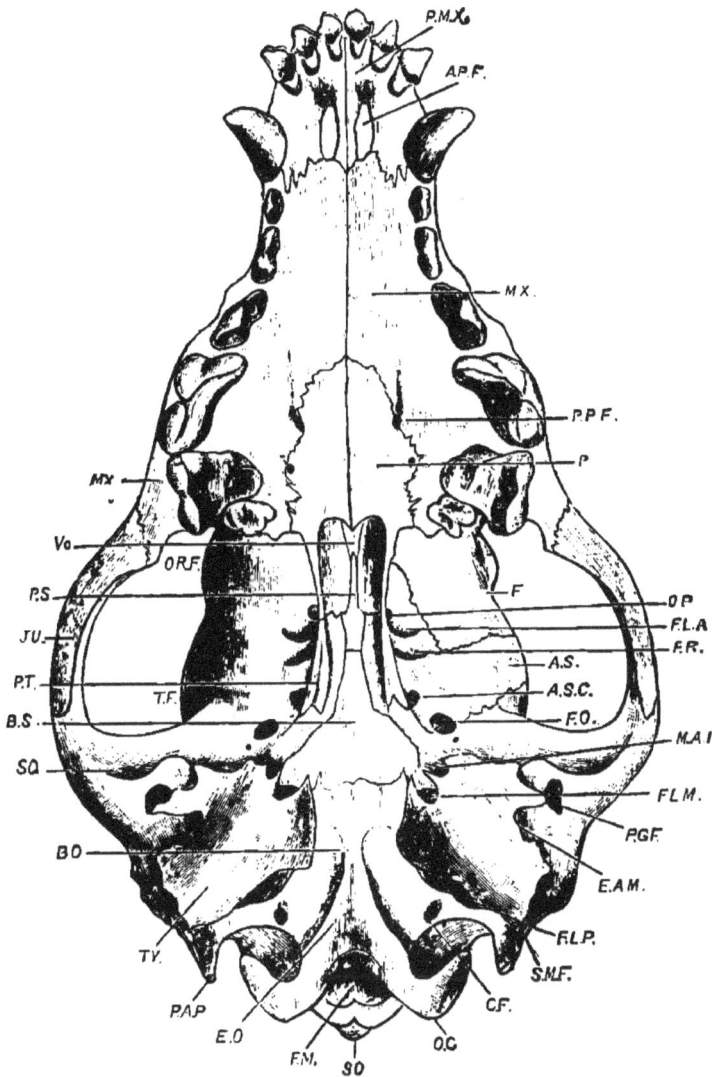

Fig. 6.—*Canis familiaris.* Ventral Surface of the Skull. (F.W.C.)

A.P.F. Anterior palatine foramen. A S. Ali-sphenoid. A S.C. Ali-sphenoid canal, posterior opening. B O. Basi-occipital. B S. Basi-sphenoid. C F. Condylar foramen. E.A.M. External auditory meatus. E O. Ex-occipital. F. Frontal. F.L.A. Foramen lacerum anterius. F.L.M. Foramen lacerum medium. F.L.P. Foramen lacerum posterius. F M. Foramen magnum. F O. Foramen ovale. F R. Foramen rotundum. M A. Malar. M.A.I. Meatus auditorius internus. M X. Maxilla. O C. Occipital condyle. O P. Optic foramen. O R.F. Orbital fossa. P, Palatine. P A.P. Par-occipital process. P G.F. Post-glenoid foramen. P S. Pre-sphenoid. P T. Pterygoid. P.MX. Pre-maxilla. S M.F. Stylomastoid foramen. S O. Supra-occipital. S Q. Squamosal. T F. Temporal fossa. T Y. Tympanic. V O. Vomer.

it transmits the main trunk of the **facial nerve.**

v. The **external auditory meatus** is the large dorso-lateral opening in the auditory bulla.

vi. The **post-glenoid foramen** is situated immediately behind the process of the squamosal; it transmits a vein from the **lateral sinus.**

vii. The **foramen lacerum medium** is an oval aperture between the posterior portion of the ali-sphenoid; through it the **internal carotid artery** passes to the cranial cavity.

viii. The **internal auditory meatus** is situated on the cranial side of the petrous portion of the periotic, and permits the passage of the **auditory nerve** (viii) to the internal ear, and of the **facial** (vii) to the stylomastoid foramen.

ix. The **foramen ovale** opens at the basal portion ot the ali-sphenoid. It transmits the third or **mandibular division** of the **fifth nerve.**

x. The **ali-sphenoid canal** opens immediately in front of the foramen ovale. It passes in a horizontal direction forwards in the ali-sphenoid, and opens into the **foramen rotundum.**

The canal is traversed by the **external carotid artery.**

xi. The **foramen rotundum** opens directly anterior to the posterior opening of the ali-sphenoid canal and transmits the **maxillary division** of the **fifth nerve.**

xii. The **foramen lacerum anterius,** or **sphenoidal fissure,** is situated between the basal portions of the ali-sphenoid and orbito-sphenoid. It is a large irregular opening and transmits the **motor-oculi** (iii), **pathetic** (iv), and **abducent** (vi), nerves, and the **ophthalmic division** of the fifth **nerve.**

xiii. The **optic foramen** is a large aperture in the orbito-sphenoid: it transmits the **optic nerve.**

xiv. The **posterior palatine foramina** are a series—usually six—of small apertures, bordering the suture between the palatines and maxillæ; they transmit branches of the **fifth nerve** and **blood vessels.**

xv. The **anterior palatine foramina** are two large oval apertures in the median line bordered posteriorly by the maxillæ, and anteriorly by the premaxillæ. Through them the **naso-palatine branch** of the **fifth nerve** passes.

xvi. The **inter-orbital foramina** perforate the anterior portion of the palatine, on the inner wall of the orbit. They transmit the **nasal branch** of the **ophthalmic division** of the fifth nerve.

xvii. The **infra-orbital foramina** are two large apertures, one situated on either side of the maxilla beneath the jugals. They transmit branches of the **maxillary division** of the fifth nerve.

xviii. A series of foramina perforate the posterior border of the ethmo-turbinal, through which branches of the **olfactory nerve** pass. This perforated plate is termed the **cribriform plate.**

xix. The **inferior dental foramen** is a large aperture on the inner side of the ascending ramus of the mandible, slightly above and in front of the angular process. It leads into the **dental canal,** and admits a branch of the **fifth nerve** known as the **dental nerve,** also an **artery.**

xx. The **mental foramen** is a small aperture on the outside of the horizontal ramus of the mandible ; and, in the anterior portion, through it a branch of the **dental nerve** makes its exit.

FORAMINA FOR THE PASSAGE OF CRANIAL NERVES.

i. **Olfactory.** Passes through numerous foramina in the cribriform plate.

ii. **Optic.** Passes through the optic foramen, a large aperture in the orbito-sphenoid.

iii. **Oculo-motor.**
iv. **Pathetic.**

Both pass through the foramen lacerum anterius, situated between the basal portions of the ali- and orbito-sphenoids.

v. **Trigeminal.**

   1. **Ophthalmic division.** Passes through the foramen lacerum anterius.

   2. **Maxillary division.** Passes through the foramen rotundum, and infra-orbital foramen.

   3 **Mandibular division.** Transmitted by the foramen ovale.

vi. **Abducent.** Passes through the foramen lacerum anterius.

vii. **Facial.** Leaves the skull by the stylomastoid foramen.

viii. **Auditory.** Passes to the internal ear through the periotic, by the internal auditory meatus.

ix. **Glossopharyngeal.**
x. **Vagus.**
xi. **Spinal Accessory.**

All make their exit from the skull by the foramen lacerum posterius.

xii. **Hypoglossal.** Passes out through the condylar foramina, situated in the ex-occipitals near the condyle.

THE TEETH.

The teeth while forming no part of the skeleton, and having quite a different origin, are so closely related to the maxillary and mandibular arches, and offer such valuable aid in the identification and classification, that they may here be conveniently described in completing the description of the dog's skull.

Development: The teeth are developed in the mucous membrane of the jaw, the enamel being produced from the epithelium *(epiblastic)*, and the dentine, pulp, and cement from the sub-epithelial connective tissue *(mesoblastic)*.

The first sign of a tooth in the embryo, is an involution of the epithelium to form a groove, the **primitive dental groove,** which is filled with the involuted epithelial cells. At the bottom of this groove, from a series of differentiated corpuscles in the sub-epithelial tissue, a **dental papilla** is formed. The involuted epithelial cells increase in number, and certain of these become modified, and form a cup over the apex of the papilla, and ultimately become the **enamel organ.** The papilla grows, and becomes differentiated into **formative pulp** and **dentine.** Calcification later takes place, successive thin layers of dentine forming as the milk teeth grow. Finally the papilla becomes narrower by the continued calcification of the pulp, leaving only a central pulp cavity, into which blood vessels and nerves pass. A portion of the involuted epithelium becomes separated off from the original sac, and a new papilla rises at its side ; from this the permanent tooth arises.

Structure. Each tooth consists of three tissues, **enamel, dentine** and **cement,** covering a soft vasculated tissue, the **pulp.**

Enamel forms the outermost part of each tooth. It is the hardest of all the tissues, containing from 95 to 97 per cent. of mineral substances (calcium phosphate and fluoride, calcium carbonate, magnesium phosphate, &c.), and the smallest amount of organic matter. It consists of a series of slender prismatic fibres.

Dentine constitutes the greater part of a tooth. Although not identical with, it is not unlike, bone. It consists of a matrix, largely impregnated with calcium phosphate, and permeated by a series of fine branched tubes—dentinal tubes—arising from the pulp cavity. In each is a soft uncalcified fibre, which is continuous with a cell on the surface of the pulp (Tomes). There are a number of varieties of dentine, known as Plici-dentine, Vaso-dentine, Osteo-dentine, &c.

Cement. The cement forms a coating over the fangs of each tooth. It is very closely allied

to bone, consisting of a laminated matrix with
lacunæ, canaliculi, and, when of considerable
thickness, vascular canals agreeing with Haversian canals.

The Pulp occupies the central portion of the tooth.
It is richly supplied with blood vessels and
nerves and consists of a gelatinous connective
tissue, containing numerous cells. It is largest
in developing teeth, often in later life becoming
converted into a form of dentine.

As the teeth are used largely by anatomists in
diagnosing the characters of the skull a "dental
formulæ" has arisen, expressing by numbers, or
letters and numbers, the nature of the dentition.
Thus we express the typical mammalian den-
tition as follows : i $\frac{3}{3}$, c $\frac{1}{1}$, p $\frac{4}{4}$, m $\frac{3}{3}=\frac{11}{11}$ total 44,
the letters i, c, p, m, standing for incisors,
canines, premolars, and molars. It being suf-
ficient to enumerate the teeth on one side of
the jaw only, it may be abbreviated to such a
formula as the following :

Man $\frac{2123}{2123}=32$ Dog $\frac{3142}{3143}=42$.

Where the incisors are separated from the
molars by a gap (as in Rodents) the space is
termed a **diastema.**

a. The teeth of the upper jaw.

i. The **incisors.** There are six in the
upper jaw, in alveoli in the premaxilla. The last—*i* 3—is much larger than
those in front of it. Each has a single

root and on its surface a groove, dividing the crown into three cusps, of which the centre one is the largest.

iii. The **premolars.** There are four pairs borne by the maxilla. They increase in size from before backwards, the last —p 4—becoming specially modified. It is known as the **sectorial** or **carnassial** tooth. It forms a blade-like surface and has three cusps, the anterior and inner one being the smallest, the middle one the highest and most pointed, and the posterior having a sharp straight ridge. There are three roots, that of the inner cusp being distinct.

ii. The **canines** are separated from the incisors by an interspace, they are two in number, borne by the maxilla. Each is slightly curved and pointed.

iv. The **molars** are the two pairs of posterior teeth. The first is the largest. They have no deciduous predecessors.

b. **The teeth of the lower jaw.**

i. The **incisors** are six in number, as in the upper jaw. They differ from them in being slightly smaller, and in having the cusps more strongly marked.

ii. The **canines** of the lower jaw are

grooved on their inner and anterior surface, and sharper on their posterior surface. They are also slightly larger than those in the upper jaw.

iii. The **premolars** are eight in number. They are more compressed, and sharper, than those of the upper jaw.

iv. The **molars.** There are three pairs of molars in the lower jaw. The first is very different from the upper one. It is the **sectorial** tooth of the lower jaw. It has four cusps.

SECTIONS.

Many points in the structure of the skull, such for instance as the relations of the bones of the occipital segment, auditory capsule, nasal cavities, &c., are best studied by an examination of a series of sections.

A. **Longitudinal Vertical Section** slightly to the left of the median line (Fig. 7).

Notice that the skull has a longitudinal central axis—the **cranio-facial axis**—the bones of which form the basal region of the skull, and support the three segments of the cranial region, which form the cranial cavity and enclose the brain. This cavity may conveniently be divided into three compartments, the

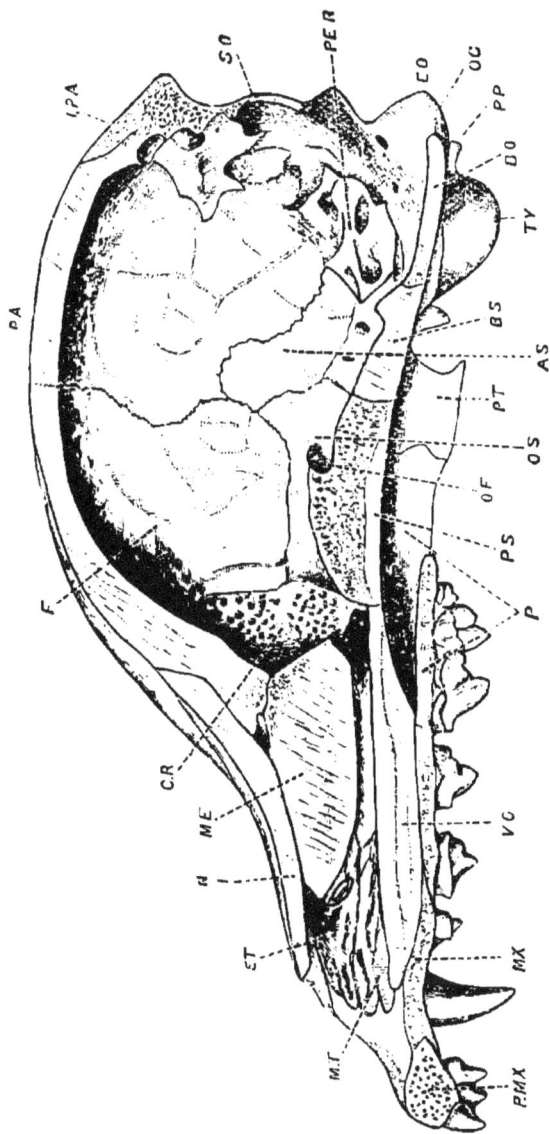

Fig. 7.—*Canis familiaris.* **Longitudinal vertical section of the Skull.** (F.W.C.)

A S. Ali-sphenoid. **B O.** Basi-occipital. **B S.** Basi-sphenoid. **C R.** Cribriform plate. **E O.** Ex-occipital. **E T.** Ethmoturbinal. **F.** Frontal. **I. P A.** Interparietal. **M E.** Mesethmoid. **M.T.** Maxillo-turbinal. **M X.** Maxilla. **N.** Nasal. **O.C.** Occipital condyle. **O.F.** Optic foramen. **O S.** Orbito-sphenoid. **P.** Palatine. **P A.** Parietal. **P E R.** Periotic. **P.M X.** Pre-maxilla. **P.P.** Par-occipital process. **P S.** Pre-sphenoid. **P T.** Pterygoid. **S O.** Supra-occipital. **T Y.** Tympanic bulla. **V O.** Vomer

E

most posterior the **cerebellar fossa**, extending as far as the periotic bone and the junction of the supra-occipital and parietal, it lodges the cerebellum ; anterior to this is the **cerebral fossa**, lodging the cerebrum. On the posterior portion of the orbito-sphenoid and frontal is a slight ridge, which forms an imperfect division, and lodges the frontal and temporal lobes of the brain respectively.

In the **occipital segment** notice the basi-occipital forming the inferior margin of the foramen magnum, and its suture with the ex-occipital; above the ex-occipital is a larger cancellous bone, the supra-occipital, dorsal to which, and between the two parietals, is a long narrow bone, the inter-parietal.

The **parietal segment** consists of the basi-sphenoid below, in which notice in the middle, on the inner surface, a little hollow—the **sella turcica** —which lodges the pituitary body. The basi-sphenoid is bounded on either side by a wing-like bone, the ali-sphenoid, dorsal to which are the large parietals. Notice that between the occipital and parietal segments there are a series of bones—the periotic and squamosal— which unite the two segments inferiorly.

The basal portion of the **frontal segment** is formed by the pre-sphenoid, from the sides of which project the orbito-sphenoids ; above these the frontals are situated, forming the dorsal, and

part of the anterior portion of the cranial
cavity.

The **facial portion** of the skull is separated from
the cranial cavity by the cribriform plate,
which is lodged between the pre-sphenoid,
orbito-sphenoids, and frontals. It is further
divided into two tube-like portions (better seen
in tranverse section), which constitute the nasal
cavities. The median partition is formed by
the mesethmoid cartilage, which is situated in a
groove of the vomer, the dorsal portion of which
is produced into a thin vertical plate and also
enters into this dividing septum, in the ventral
portion. The sides of the vomers are laterally
produced and form a partial division between
the nasal cavity and the olfactory chamber (see
transverse section D). The anterior and posterior portions of the nasal cavities are filled
by a series of infolded laminæ of bone—the
maxillo-turbinals and ethmo-turbinals; inferiorly are a similar series—the naso-turbinals.
Enclosing the nasal cavities and olfactory
chamber, and forming the outer wall of the
facial portion of the skull, are the nasals,
maxilla, premaxilla, palatines, and pterygoids.
Viewed in longitudinal section, only a small
portion of these bones are visible. The nasals
form the dorsal portion of the nasal cavities,
and articulate with the frontals; they are situated immediately above the ossified mesethmoid

cartilage. The anterior portion of the middle
of the maxilla is visible as a thin vertical
plate ; in front of which is the premaxilla, its
inner border forming the anterior narial

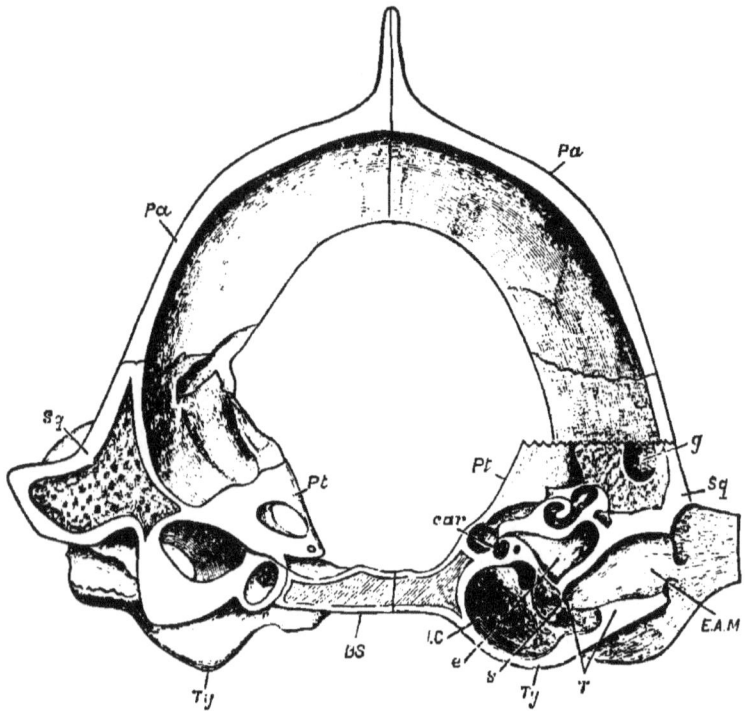

Fig. 8.—Transverse section through the posterior portion
of the Skull, showing the tympanic bulla
in section. (W.E.C. and F.W.C.)

B S. Basi-sphenoid. **Car.** Carotid Canal. **E.** Eustachian canal. **E.A.M.**
External auditory meatus.   **I C.** Inner chamber.   **P A.** Parietal.   **Pt.**
Periotic. **Sq.** Squamosal. **S.** Septum. **T.** Tympanic ring. **Ty.** Tympanic.

opening. A portion of the palatine, immediately behind the maxilla, forms a vertical plate,
its edge rising and meeting with the vomer.

A thin vertical plate of bone, articulating anteriorly with the palatine, and lying beneath
the pre-sphenoid and orbito-sphenoid, is the
pterygoid.

B. **Transverse Section through the posterior
portion of the Skull.** (Fig. 8).

In the left hand of the section the periotic is shown
lying dorsal to the tympanic bulla. Notice on
the right hand side, that a portion of the
anterior wall of the tympanic bulla has been
cut through, and laterally the external auditory
meatus. Dorsally, the prominent sagittal crest
forms a median prominence. The parietals
form the dorso-lateral boundary of the section,
the ventral and lateral portions being com-
pleted by the squamosals. Lying on the inner
border of the squamosal, notice the periotic.

In the section of the bulla, notice that the cavity
is imperfectly divided into two chambers by
an incomplete **bony septum,** arising from the
anterior wall. The inner chamber terminates
blindly ; while into the outer, the Eustachian
tube enters. The somewhat prominent lip of
the external auditory meatus, is well seen in
section. On the inner side of the section,
notice the cut portion of the carotid canal.

C. **Transverse Section passing in front of the
fronto-parietal suture, and the anterior
border of the pre-sphenoid.** (Fig. 9).

The dorsal portion of the section is formed by the

frontals; on the sides of which, notice the post-
orbital processes.

The cribriform plate is bounded dorsally and laterally by the frontals; beneath and posterior to
the cribriform plate, the most anterior portion
of the pre-sphenoid is visible, and anteriorly
the vomer.

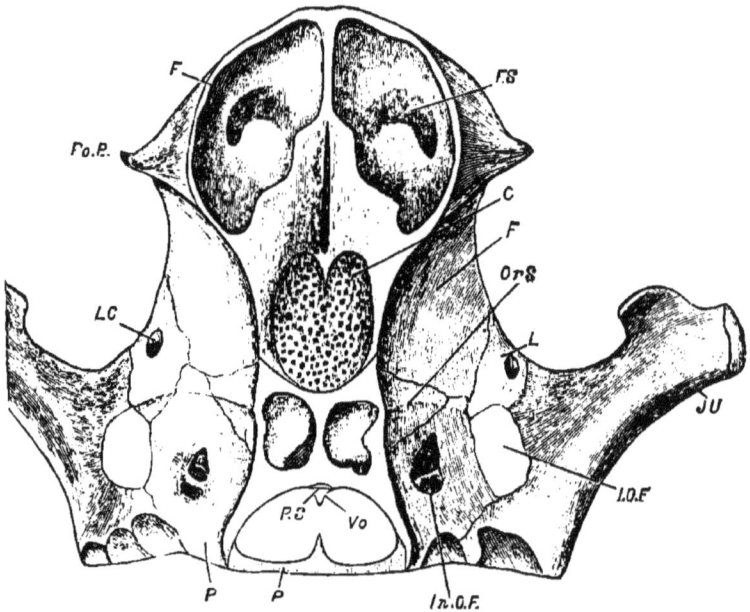

Fig 9.—Transverse section through the Skull passing in
front of the fronto-parietal suture and the
anterior border of the Pre-sphenoid. (F.W.C.)

C. Cribriform plate.   F. Frontal.   F S. Frontal Sinus.   In.O.F. Inter-
orbital foramen.   I.O.F. Infra-orbital foramen.   J U. Malar.   L. Lachrymal.
L.C. Lachrymal canal.   Or S. Orbito-sphenoid.   P. Palatine.   Po.P. Post-
orbital process.   P S. Pre-sphenoid.   Vo. Vomer.

Laterally, notice the lachrymals, perforated by the
lachrymal canal.

The malars form the most lateral portion of the
section, and bound the infra-orbital foramen.
The orbito-sphenoid, the anterior portion, is visible
lying between the frontal and palatine.
The palatines form the floor and the lateral portions

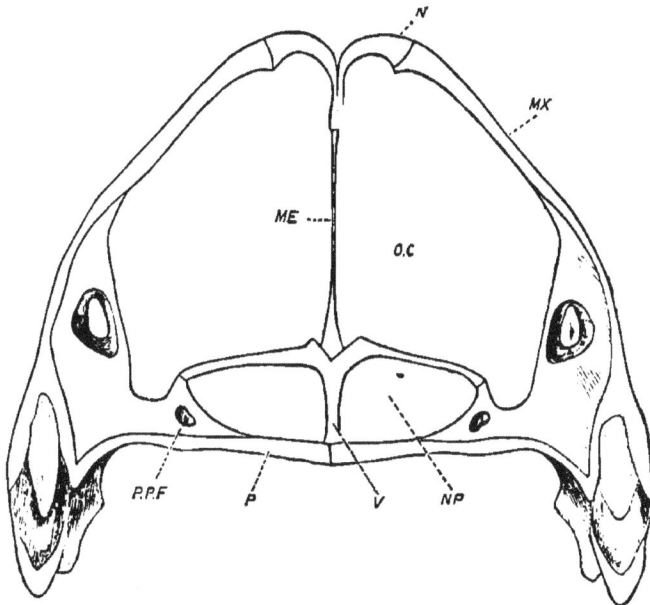

Fig. 10.—Transverse section anterior to the cribriform
plate. (W. F. C.)

M E. Mesethmoid. M X. Maxilla. N. Nasal. N.P. Nasal passage.
O.C. Olfactory chamber. P. Palatine. P.P.F. Posterior palatine foramen.

of the section ; beneath the orbito-sphenoid
they are perforated by the inter-orbital foramen.
D. **Transverse Section through the skull, anterior to the cribriform plate.** (Fig. 10).
The section is bounded dorsally by the nasals and
maxillæ, and ventrally by the palatines. The

mesethmoid is a thin plate-like bone, forming a median partition between the olfactory chambers ; dorsally it is attached to a **ventral process** of the nasals.[1] The ventral border of the mesethmoid is thickened and lodged in the dorsal portion of the vomer. The vomer, in the portion shown in section, forms a complete

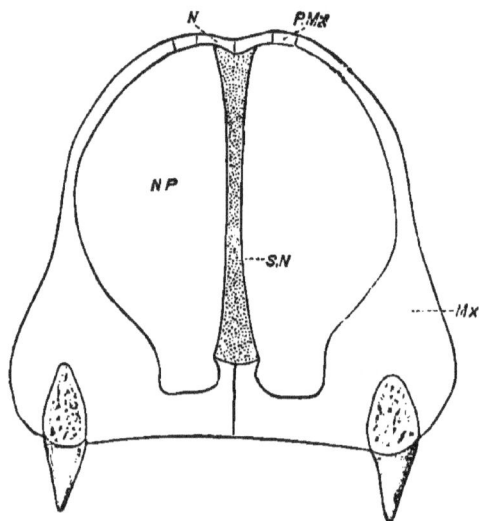

Fig. 11.—**Transverse section through the anterior portion of the nares.** (W.F.C.)

M X. Maxilla. N. Nasal. N P. Nasal Passage. P.M X. Pre-maxilla.
S.N. Septum nasi.

septum between the right and left nasal passages, by meeting ventrally with the palatines. Dorsally it is trough-like, or V-shaped ; the two lateral arms embracing the ventral border of the mesethmoid. From these lateral arms,

[1] In old skulls the nasals fuse, and this process is very distinct.

wing-like processes extend to the right and left of the nasal cavity, and form a horizontal partition, separating the nasal passages from the olfactory chambers. The horizontal laminæ of the palatines complete the floor of the nasal passages. In the transverse section E, through the anterior region of the maxillæ, the section is bounded dorsally by the nasals, and the posterior portion of the dorsal part of the premaxillæ; the sides and floor are formed by the maxillæ. Ventrally the maxillæ form a horizontal plate, anterior to which the floor is formed by the pre-maxillæ.

REFERENCES.

FLOWER, W. H. & GADOW, H.
"An Introduction to the Osteology of the Mammalia." (Skull, pp. 116-242.) London, 1885.

HUXLEY, T. H.
"A Manual of the Anatomy of Vertebrated Animals." (Dog's Skull, pp. 353-4.) London, 1871.

MARSHALL, A. MILNES, & HURST, C. HERBERT.
"A Junior Course of Practical Zoology." (Rabbit's Skull, pp. 294-313.) London, 1895.

MIVART, ST. GEORGE.
"The Cat." London, 1881.

MORRELL, G. H.
"The Student's Handbook of Comparative Anatomy, Pt. I. Mammalia." (Sheep's Skull, pp. 208-260.) London, 1872.

OWEN, R.
"The Anatomy of Vertebrates." (Vol. II. p.
297.) London, 1866.

PARKER, W. K. & BETTANY, G. T.
"The Morphology of the Skull." (Skull of the
Pig, &c., pp. 267-309.) London, 1877.

PARKER, T. JEFFREY.
"A Course of Instruction in Zootomy." (Rabbit's Skull, pp. 268-278.) London, 1884.

WEIDERSHEIM, R.
"Lehrbuch, dor Vergleichenden Anatomie der
Wirbelthiere." (Skull, pp. 151-161.)
Jena, 1886.

WEIDERSHEIM, R. & PARKER, W. NEWTON.
"Elements of the Comparative Anatomy of
Vertebrates." (Skull, pp. 80-84.) Lon-
don, 1886.

CHAPTER IV.

GENERAL REVIEW OF THE SKULLS OF THE CARNIVORA, AND THE DIFFERENT BREEDS OF DOGS.

The term Carnivora, when first used, included
the Insectivora and Cheiroptera, as well as the Carnivora, as restricted by Huxley, Owen, and Flower.
For this reason, certain authors have suggested and
used the term Sectorialia, which, however, has not
been generally adopted.

ORDER CARNIVORA.

Sub-order FISSIPEDIA.

Comprises the Cats, Dogs, and Bears. Cuvier divided the sub-order into two groups :

Plantigrada, walking on the sole of the foot; and *Digitigrada*, walking only on the toes.

The distinction is not one of any great importance, seeing that almost every intermediate condition exists.

Section ÆLUROIDEA. Comprises the cat-like forms, and includes the following families :

1. **Felidæ.** Skull usually short, especially facial portion. In *F. uncia* (the Ounce) the facial portion is much broader than in any other member of the family, the nasal bones being broad and flat. Zygomatic arch prominent. Mastoid process often absent, never very conspicuous. Auditory bulla smooth, rounded, and large ; with an almost complete bony septum between the two chambers of the bulla. There is no ali-sphenoid canal; and the canal for the passage of the carotid artery is very small. Condylar and glenoid foramina small. An interesting fossil (Eocene), form *Eusmilus*, grouped under this family, has four incisor teeth in the lower jaw, and small canines, which are separated from a single premolar and true molar by a diastema.

2. **Viverridæ.** Includes such forms as the Civet-Cats, Ichneumons. Skull more elongated in the facial

portion than in the *Felidæ*. Carotid canal distinct, running as a groove on the side of the bulla. Auditory bulla ossified (ex. *Nandinia*), large, and divided by a septum. In the sub-family *Herpestinæ*, it is somewhat pyriform in shape, the posterior chamber being large, and rounded. Ali-sphenoid canal generally present (ex. *Cryptoproctinæ, Herpestinæ, Galidictinæ, Euplerininæ*). In *Eupleres*, of which genus only one species is known, the *E. goudoti* of Madagascar, the jaws are slender and of small size, and there is no ali-sphenoid canal.

3. **Proteleidæ.** This genus contains but a single species, *Proteles cristatus*. The molar teeth are rudimentary in character, being small, and placed far apart. There is no ali-sphenoid canal. The auditory bulla consists of two separate chambers.

4. **Hyenidæ.** Skull with no ali-sphenoid canal. Sagittal crest prominent. Auditory bulla not divided, but rudiment of septum may be present, and usually anchylosed into one mass.

Section CYNOIDEA. Contains only one family.

1. **Canidæ.** Includes the Dogs, Wolves, Jackals, and Foxes. The Dog's skull may be regarded as typical of the family.

Section ARCTOIDEA. Comprises the Bears, Otters, Badger, and Weasels.

1. **Ursidæ.** Facial portion short. Zygomatic arches and sagittal crest largely developed (especially

so in *Æluropus*). Auditory bulla depressed. Orbits small, and incomplete posteriorly. Paroccipital process prominent. Condylar foramen exposed. Excepting in *Æluropus*, there is an ali-sphenoid canal.

2. **Procyonidæ.** Facial portion of the skull usually short and broad, cranium high and compressed. Zygomatic arches strong. Coronoid process of mandible strongly developed.

3. **Mustellidæ.** Cranium elongated and narrow in many genera. The post-glenoid process is often produced over the deep glenoid fossa (e.g. *Meles*), and fits lightly over the condyle of the mandible. No ali-sphenoid canal. Dentition variable ; in *Latax* there are but two incisors in each side of the lower jaw, thus differing from all other Carnivora. The paroccipital process of the ex-occipital usually distinct from the bulla, triangular in shape, and directed outwards, downwards, and backwards.

Sub-order ii PINNIPEDIA.

The families of this sub-order are : 1. **Otariidæ ;** 2. **Trichechidæ ;** 3. **Phocidæ.** They are all aquatic, and differ from the rest of the Carnivora in the modification of the limbs for progression in water. The cranial characters have been tabulated by Mivart as follows :

" 1. No complete septum in the auditory bulla of any genus.

2. The lip of the meatus auditorius externus pro-

jects greatly outwards in the *Phocidæ;* but it is not the median inferior part of the lip, as in the Bears; but posteriorly, as in the Otters. It is not prolonged outwards in the *Otariidæ* and *Trichechidæ.*

3. The paroccipital process is more or less triangular, and directed outwards, downwards, and backwards, except in *Trichechus*, where it forms a small buttress against, and uniting with, the hinder side of the great mastoid.

4. The mastoid process may be considerably prominent (as in *Otaria*), or extremely so (as in *Trichechus*), or may form part of a wide-spread rounded prominence (as in the *Phocidæ*). It may form a continuous bone wall with the paroccipital process (as in *Otaria*), or be separated from it (as in the *Phocidæ*), or blend with it (as in *Trichechus*).

5. The carotid foramen is always large and conspicuous; and is placed towards, or almost at the hinder end of, the bulla, which the carotid canal traverses, towards or along its inner margin — its course being indicated externally in *Otaria* and *Trichechus*, but not at all in the *Phocidæ*. It is never concealed (as it is in the Bears) by a projecting lip of the basi-occipital.

6. The condyloid foramen is always distinct and exposed, and never overlapped by a ridge of bone running from the paroccipital process of the condyle; and never opens into, though it appears sometimes to

coalesce with, the foramen lacerum posterius.

7. The glenoid foramen is always very small, and is sometimes not to be detected.

8. The alisphenoid canal may be present or absent.

9. The suborbital foramen is always rather large ; but never as large relatively as in *Lutra* and *Enhydra*. It is largest in *Trichechus*.

10. The frontal postorbital process, present in *Otaria* and *Trichechus*, is never more than a rudiment in the *Phocidæ*.

11. The zygomatic postorbital process is formed both by the malar and squamosal in the *Phocidæ*, mainly by, the malar in the *Otaria*, and entirely by it in *Trichechus*.

12. The ali-sphenoid and parietal always join by a narrow process of the latter bone.

13. The premaxillæ never ascend to join the frontals.

14. There is never a lachrymal foramen.

15. The basis cranii is nearly always bent, so as to be convex downwards.

16. The anterior nares are quite terminal in *Trichechus*; rather more distant from the end of the muzzle in *Otaria;* and not at all terminal, but looking more or less exteriorly, upwards as well as forwards, in the *Phocidæ*.

17. The opening represents both the foramen rotundum, and the spheno-orbital fissure.

18. The optic foramina open into the cranial cavity by a single aperture in *Otaria* and in *Stenorhynchus*, but not in the *Phocidæ* generally, as in *Trichechus*.

19. The palate always extends backwards, much behind the last molars ; but is not commonly narrowed behind, save in *Otaria*. It is not at all so narrowed in *Trichechus*.

20. Defects of ossification commonly occur in the occipital in the *Phocida*, but not in *Otaria* and *Trichechus*.

21. A preorbital process exists in *Otaria* and *Trichechus*; sometimes, but rarely, in the *Phocidæ*.

22. The angle of the mandible is inflected (as in Marsupials) in *Otaria*, but not in the other genera."

THE CRANIAL AND DENTAL CHARACTERS OF THE DIFFERENT BREEDS OF DOGS.

The tables i and ii detail a series of measurements, made on the skulls of a number of different breeds of dogs. In table i the maximum and minimum measurement of each breed is given, the difference between these two, and the average. It will be seen that the averages in the different breeds differ very little from one another ; this is perhaps more noticeable, in table ii, in the teeth. In table iii the length and breadth are compared.

Table I. COMPARATIVE MEASUREMENTS.—CRANIAL.

	TOTAL LENGTH.			
	Max.	Min.	Diff.	Av.
Esquimaux	339·09	269·23	69·86	299·08
Sheep-dog	300·78	261·66	39·12	288·50
Newfoundland	321·62	297·14	24·48	308·01
Greyhound	359·70	280·00	79·70	313·74
Italian Greyhound	275·05	268·00	7·05	271·29
Irish Wolf-dog, modern	337·41	304·34	33·07	319·18
Irish Wolf-dog, old	326·66	284·84	42·82	304·47
Spaniel	313·79	258·18	55·61	283·19
Bloodhound	348·33	284·70	63·63	307·59
Pointer	300·0	252·99	47·01	280·08
Mastiff	356·16	279·86	76·30	306·51
Bull-dog	350·00	244·80	105·20	278·41
Pug	238·88	213·63	25·25	227·40
Fox-Terrier	322·22	260·41	61·81	284·61
Skye-Terrier	320·23	272·72	57·51	303·61
Pariah	320·32	284·48	35·84	301·64
Dingo	323·30	283·60	39·70	302·57

Table I.—COMPARATIVE MEASUREMENTS. — CRANIAL.

| | ZYGOMATIC WIDTH. | | | |
	Max.	Min.	Diff.	Av.
Esquimaux	197·81	156·41	41·40	170·62
Sheep-dog	173·43	136·66	37·77	157·01
Newfoundland	189·18	157·14	32·04	173·43
Greyhound	167·18	138·46	29·72	151·30
Italian Greyhound	164·00	145·83	18·17	155·82
Irish Wolf-dog, modern	169·56	152·63	16·93	162·03
Irish Wolf-dog, old	176·66	153·73	23·93	165·70
Spaniel	175·86	154·54	21·32	163·82
Bloodhound	191·66	154·28	37·38	166·96
Pointer	175·92	148·68	27·24	166·02
Mastiff	203·38	162·33	41·05	175·73
Bull-dog	238·72	171·42	67·30	201·21
Pug	188·63	157·14	31·49	168·96
Fox-Terrier	195·55	169·82	26·73	178·57
Skye-Terrier	210·52	172·72	37·80	193·64
Pariah	188·18	149·19	38·99	165·75
Dingo	172·22	167·74	9·48	172·71

Table I.—COMPARATIVE MEASUREMENTS.—CRANIAL.

LENGTH OF BONY PALATE.

	Max.	Min.	Diff.	Av.
Esquimaux	162·96	138·46	24·50	145·71
Sheep-dog	150·86	130·83	20·03	141·26
Newfoundland	156·75	146·60	10·15	150·12
Greyhound	153·96	147·76	6·20	150·35
Italian Greyhound	143·52	135·41	8·11	139·64
Irish Wolf-dog, modern	161·29	146·05	15·24	153·65
Irish Wolf-dog, old	153·33	133·33	20·00	143·17
Spaniel	151·72	124·19	27·53	139·23
Bloodhound	171·66	142·35	29·31	150·56
Pointer	155·55	128·70	26·85	144·99
Mastiff	169·86	129·87	39·99	149·60
Bull-dog	139·36	104·00	35·36	119·67
Pug	122·22	110·71	11·51	115·52
Fox-Terrier	155·55	133·33	22·22	141·03
Skye-Terrier	165·11	140·90	24·21	152·88
Pariah	156·09	127·58	28·41	146·77
Dingo	156·60	141·93	14·67	148·92

68

Table I.—Comparative Measurements. Cranial.

WIDTH OF BONY PALATE.

	Max.	Min.	Diff.	Av.
Esquimaux	119·09	92·30	27·79	109·38
Sheep-dog	109·44	91·66	17·78	99·89
Newfoundland	113·51	100·00	13·51	105·38
Greyhound	93·75	87·69	6·06	90·78
Italian Greyhound	100·00	89·65	10·35	94·46
Irish Wolf-dog, modern	103·22	96·05	7·17	100·24
Irish Wolf-dog, old	105·00	88·60	17·40	96·36
Spaniel	112·06	100·00	12·06	104·81
Bloodhound	113·33	90·00	23·33	101·71
Pointer	111·11	95·72	15·39	101·93
Mastiff	115·38	95·71	19·67	104·82
Bull-dog	153·19	103·17	50·02	128·84
Pug	125·00	116·66	8·44	121·46
Fox-Terrier	117·77	103·44	14·33	108·48
Skye-Terrier	134·88	102·27	32·61	119·40
Pariah	110·68	90·32	20·36	100·34
Dingo	108·19	98·48	9·61	104·41

Table I.—COMPARATIVE MEASUREMENTS.—CRANIAL.

LENGTH OF Pm. & M.

	Max.	Min.	Diff.	Av.
Esquimaux	121·48	93·07	28·41	101·55
Sheep-dog	154·68	95·83	58·85	112·93
Newfoundland	100·00	96·00	4·00	98·32
Greyhound	111·71	104·28	7·53	107·51
Italian Greyhound	105·65	98·39	7·26	102·68
Irish Wolf-dog, modern	111·57	105·79	6·78	108·24
Irish Wolf-dog, old	109·72	96·96	13·76	104·09
Spaniel	113·20	89·09	24·11	97·37
Bloodhound	106·66	90·00	16·66	98·52
Pointer	101·58	96·47	5·11	98·70
Mastiff	110·95	84·41	26·54	99·02
Bull-dog	99·28	78·72	21·56	87·52
Pug	88·09	68·18	19·91	80·79
Fox-Terrier	102·85	95·55	7·30	98·32
Skye-Terrier	115·78	100·00	15·78	107·58
Pariah	124·78	88·70	36·08	107·63
Dingo	116·33	98·78	17·55	107·59

Table I. -COMPARATIVE MEASUREMENTS.--CRANIAL.

	LENGTH OF Pm. & M.			
	Max.	Min.	Diff.	Av.
Esquimaux	137·62	92·30	45·32	112·91
Sheep-dog	128·12	109·16	18·96	116·86
Newfoundland	142·85	109·30	33·55	129·15
Greyhound	128·57	113·49	15·06	119·27
Italian Greyhound	120·00	112·00	8·00	114·83
Irish Wolf-dog, modern	126·86	121·05	5·81	124·79
Irish Wolf-dog, old				
Spaniel	129·34	103·22	26·02	113·78
Bloodhound	119·16	105·80	13·36	114·17
Pointer	116·23	109·85	6·38	112·34
Mastiff	122·03	88·31	33·72	108·09
Bull-dog	140·00	108·30	31·70	117·23
Pug	111·90	88·63	23·37	102·49
Fox-Terrier	116·19	114·58	2·61	115·45
Skye-Terrier	129·06	120·45	8·55	125·27
Pariah	133·94	106·45	27·49	122·79
Dingo	140·00	121·05	18·95	126·32

Table II.—Comparative Measurements. —Dental.

	LENGTH OF Pm. 4.			
	Max.	Min.	Diff.	Av.
Esquimaux	28·72	25·64	13·08	30·74
Sheep-dog	30·17	27·50	2·67	28·83
Newfoundland	28·57	27·70	·87	28·27
Greyhound	29·23	26·26	2·97	28·13
Italian Greyhound	31·29	28·00	3·29	29·48
Irish Wolf-dog, modern	29·71	25·80	3·91	27·27
Irish Wolf-dog, old	30·83	25·75	5·08	28·93
Spaniel	33·33	24·13	9·20	29·30
Bloodhound	30·00	23·52	6·48	26·94
Pointer	30·95	24·64	6.31	28·29
Mastiff	31·61	24·02	7·59	27·22
Bull-dog	31·91	26·00	5·91	28·47
Pug	31·11	28·57	2·54	29·74
Fox-Terrier	34·44	27·58	6·86	30·72
Skye-Terrier	39·67	31·81	7·86	36·11
Pariah	35·57	28·22	7·35	31·09
Dingo	33·63	29·06	4·54	30·58

Table II.—COMPARATIVE MEASUREMENTS.—DENTAL.

	LENGTH OF M. 1.			
	Max.	Min.	Diff.	Av.
Esquimaux	24·36	16·06	8·30	20·53
Sheep-dog	20·43	19·82	1·61	20·60
Newfoundland	20·71	20·00	·71	20·41
Greyhound	21·53	18·25	3·28	20·10
Italian Greyhound	23·52	20·00	3·52	21·45
Irish Wolf-dog, modern	21·73	19·61	2·12	20·36
Irish Wolf-dog, old	19·44	13·63	5·81	15·83
Spaniel	22·72	20·40	2·32	21·56
Bloodhound	21·16	17·85	3·31	19·01
Pointer	22·22	20·15	2·07	19·99
Mastiff	21·32	16·88	4·44	18·88
Bull-dog	25·95	19·84	6·11	21·77
Pug				19·20
Fox-Terrier	24·44	20·83	3·61	22·29
Skye-Terrier	26·31	20·45	5·86	22·95
Pariah	21·65	16·93	4·72	20·12
Dingo	23·66	18·06	5·60	20·37

Table II.—COMPARATIVE MEASUREMENTS.—DENTAL.

	BREADTH OF M. I.			
	Max.	Min.	Diff.	Av.
Esquimaux	30·90	22·05	8·85	26·32
Sheep-dog	27·34	24·40	2·94	26·01
Newfoundland	25·00	22·29	2·71	24·04
Greyhound	26·15	23·30	3·12	24·95
Italian Greyhound	27·05	24·00	3·05	25·35
Irish Wolf-dog, modern	26·81	23·68	3·13	25·09
Irish Wolf-dog, old	26·66	22·38	4·28	24·69
Spaniel	28·88	25·71	3·17	26·62
Bloodhound	26·25	21·42	4·83	23·83
Pointer	28·88	22·30	6·58	24·33
Mastiff	27·96	19·74	8·22	23·95
Bull-dog	35·10	24·5	11·50	27·24
Pug				26·78
Fox-Terrier	28·88	24·13	4·75	26·71
Skye-Terrier	26·74	25·00	1·74	26·02
Pariah	27·93	23·38	4·55	26·38
Dingo	30·33	25·08	2·25	26·54

Table II.—COMPARATIVE MEASUREMENTS.—DENTAL.

	LENGTH OF $\frac{M. 2.}{}$			
	Max.	Min.	Diff.	Av.
Esquimaux	13·48	8·97	4·51	10·93
Sheep-dog	14·17	9·48	4·69	11·88
Newfoundland	12·60	10·81	1·73	11·74
Greyhound	11·93	10·19	1·74	11·40
Italian Greyhound	12·00	10·56	1·44	11·34
Irish Wolf-dog, modern	12·75	11·61	1·14	12·07
Irish Wolf-dog, old	12·5	9·09	3·41	11·35
Spaniel	13·33	10·20	3·13	11·58
Bloodhound	12·50	10·00	2·50	11·10
Pointer	13·88	9·57	4·31	11·10
Mastiff	12·71	9·72	2·99	11·06
Bull-dog	14·89	9·60	5·29	12·11
Pug				9·05
Fox-Terrier	10·85	10·34	·51	10·53
Skye-Terrier	13·15	10·22	2·93	12·05
Pariah	13·28	9·67	3·61	11·48
Dingo	12·96	8·77	4·19	11·39

Table II.—Comparative Measurements.—Dental.

	BREADTH OF M. 2.			
	Max.	Min.	Diff.	Av.
Esquimaux	20·36	14·10	6·26	16·39
Sheep-dog	18·38	15·00	3·38	16·35
Newfoundland	18·00	14·86	3·14	15·89
Greyhound	17·18	14·92	2·26	16·10
Italian Greyhound	16·00	14·56	1·44	15·39
Irish Wolf-dog, modern	17·41	16·44	·97	16·93
Irish Wolf-dog, old	19·16	16·25	2·91	17·88
Spaniel	19·00	14·69	4·31	16·95
Bloodhound	17·95	13·76	4·19	17·67
Pointer	18·51	13·85	4.66	15·50
Mastiff	18·64	12·77	5·87	15·71
Bull-dog	21·27	14·16	7·11	17·05
Pug				12·84
Fox-Terrier	16·37	15·61	·76	15·86
Skye-Terrier	19·73	11·36	8·37	16·33
Pariah	17·88	13·70	4·18	16·41
Dingo	18·60	14·91	3·69	17·18

Table II.—Comparative Measurements.—Dental.

| | LENGTH OF M. I. | | | |
	Max.	Min.	Diff.	Av.
Esquimaux	39·27	29·48	9·79	34·23
Sheep-dog	34·51	31·03	3·48	32·68
Newfoundland	34·96	30·67	4·29	32·48
Greyhound	33·57	31·74	1·83	32·49
Italian Greyhound	38·35	34·00	4·35	35·92
Irish Wolf-dog, modern	34·05	31·57	2·48	33·06
Irish Wolf-dog, old				
Spaniel	37·77	32·38	5·39	35·17
Bloodhound	37·00	31·42	5·58	33·27
Pointer	35·18	30·42	4·76	32·15
Mastiff	36·76	27·08	9·68	31·51
Bull-dog	42·55	31·74	10·81	35·19
Pug	33·33	29·76	3·57	31·26
Fox-Terrier	37·77	33·33	4·44	35·03
Skye-Terrier	40·78	36·26	4·52	39·24
Pariah	39·13	30·90	8·23	34·18
Dingo	38·00	31·45	6·55	34·03

Table II.—Comparative Measurements.—Dental.

LENGTH OF $\overline{\text{M. 2.}}$

	Max.	Min.	Diff.	Av.
Esquimaux	15·71	12·05	3·66	13·86
Sheep-dog	17·32	11·70	5·62	14·11
Newfoundland	14·60	12·85	1·15	13·81
Greyhound	14·28	12·69	1·59	13·66
Italian Greyhound	14·00	12·70	1·30	13·41
Irish Wolf-dog, modern	14·47	13·76	·71	14.05
Irish Wolf-dog, old				
Spaniel	19·35	14·13	5·22	15·81
Bloodhound	14·16	10·58	3·58	12·95
Pointer	16·66	12·30	4·36	14·31
Mastiff	15·64	11·11	4·53	13·59
Bull-dog	19·04	13·6	5·54	15·56
Pug	13·88	11·90	1·98	12·76
Fox-Terrier	15·55	12·50	3·05	14·02
Skye-Terrier	15·90	13·15	2·75	14·88
Pariah	16·51	12·69	3·82	12·69
Dingo	15·00	12·29	2·71	13·24

Table II.—COMPARATIVE MEASUREMENTS.—DENTAL.

	LENGTH OF M. 3.			
	Max.	Min.	Diff.	Av.
Esquimaux	9·52	6·15	3·37	7·19
Sheep-dog	11·81	8·33	3·48	9·51
Newfoundland	9·3	6·08	3·22	7·51
Greyhound	9·37	7·69	1·68	8·29
Italian Greyhound				5·00
Irish Wolf-dog, modern				5·92
Irish Wolf-dog, old				
Spaniel	10·37	5·64	4·73	7·64
Bloodhound	9·16	5·71	3·45	7·42
Pointer	10·00	5·63	4·37	7·82
Mastiff	8·97	5·88	3·09	7·29
Bull-dog	10·00	5·92	4·8	8·26
Pug				4·54
Fox-Terrier	7·75	5·14	2·61	6·38
Skye-Terrier				
Pariah	10·09	7·14	2·95	8·58
Dingo	8·57	7·89	·68	8·26

Table III.—COMPARISON OF LENGTH AND BREADTH.

No.	TOTAL LENGTH AND ZYGOMATIC WIDTH. AVERAGE 63·53.	
	DOG.	Index.
1.	Chinese Pug-nosed Spaniel	91·13
2.	Pug	74·83
3.	Bull-dog	73·72
4.	Black-and-tan Toy Terrier	73·52
5.	King Charles	72·00
6.	Skye	63·57
7.	Fox-Terrier	62·94
8.	Turnspit	61·65
9.	English Terrier	61·53
10.	Beagle	61·07
11.	Black-and-tan Terrier	60·71
12.	Otter-dog	60·00
13.	Pomeranian	59·54
14.	Pointer	59·19
15.	Harrier	58·49
16.	Spaniel	57·78
17.	Italian Greyhound	57·63
18.	Mastiff	57·16
19.	Dingo	57·13
20.	Esquimaux	56·86
21.	Newfoundland	56·35
22.	Fox-hound	55·70
23.	Sheep-dog	55·45
24.	Bloodhound	54·79
25.	Pariah	54·28
26.	New-Zealand dog	53·57
27.	Irish Wolf-dog, old	53·02
28.	West-Indian dog	52·85
29.	St. Bernard	52·62
30.	Irish Wolf-dog, modern	50·59
31.	Greyhound	49·89

Table III.—Comparison of Length and Breadth.

No.	PALATINE LENGTH AND WIDTH. AVERAGE 76·93.	
	DOG.	Index.
1.	Chinese Pug-nosed Spaniel	120·00
2.	Bull-dog	106·84
3.	Pug	105·76
4.	King Charles	91·20
5.	Black-and-tan Toy Terrier	86·56
6.	English Terrier	81·82
7.	Harrier	80·00
8.	Skye	77·98
9.	Pomeranian	77·37
10.	West-Indian dog	76·92
11.	Fox-Terrier	76·92
12.	Spaniel	76·68
13.	Beagle	76·33
14.	Black-and-tan Terrier	72·99
15.	Otter-dog	72·36
16.	Mastiff	71·86
17.	Esquimaux	71·76
18.	Sheep-dog	70·78
19.	Newfoundland	70·23
20.	Pointer	70·21
21.	Dingo	70·03
22.	Turnspit	68·57
23.	Fox-hound	68·42
24.	New-Zealand dog	68·23
25.	Pariah	68·22
26.	Italian Greyhound	68·14
27.	Bloodhound	67·77
28.	Irish Wolf-dog, old	67·06
29.	St. Bernard	65·35
30.	Irish Wolf-dog, modern	65·17
31.	Greyhound	60·36

REFERENCES.

ALLEN, J. A.
"History of North American Pinnipeds."
Washington, 1880.
"On the Eared Seals (Otariadæ)." Bull. Mus.
Comp. Zool., Camb. U.S.A., 1870, vol. ii.
pp. 1-89, pts. 1-3.

DORAN, ALBAN H. G.
"Morphology of the Mammalian Ossecula
auditûs." Trans. Linn. Soc. (Zool.) 2nd
ser. 1879, vol. 1, pp. 371-497, pls.
lviii-lxiv.

FLOWER, W. H.
"On the Value of the Characters of the Base
of the Cranium in the Classification of
the Order Carnivora." Proc. Zool. Soc.,
1869, pp. 5-37.
Article "Mammalia." Ency. Brit., 9th ed. vol.
xv, pp. 347-446.

FLOWER, W. H. & LYDEKKER, R.
"An Introduction to the study of Mammals,
living and extinct." London, 1891.

HUXLEY, T. H.
"On the Cranial and Dental Characters of the
Canidæ." Proc. Zool. Soc., 1880, pp.
238-88.

MIVART, ST. GEORGE.
"The Cat, An Introduction to the study of
Backboned Animals, especially Mam-
mals." London, 1881.
"On the Classification and Distribution of the
Æluroidea." Proc. Zool. Soc., 1882, pp.
135-208.

"On the Anatomy, Classification and Distribu-
   tion of the Æluroidea." Proc. Zool.
   Soc., 1885, pp. 340-404.
"Notes on the Pinnipedia." Proc. Zool. Soc.,
   1885, pp. 484-500.
"A Monograph of the Canidæ." London, 1890.
TURNER, H. N.
"Observations relating to some of the Foramina
   in the Base of the Skull in Mammalia."
   Proc. Zool. Soc., 1848, pp. 63-88.
WATERHOUSE, G. R.
"On the Skulls and Dentition of the Carni-
   vora." P. Z. S., 1839, pp. 135-7.
WINDLE, B. C. A., & HUMPHREYS, JOHN.
"On Some Cranial and Dental Characters of
   the Domestic Dog." Proc. Zool. Soc.,
   1890, pp. 5-29.

CHAPTER V.

GLOSSARY OF OSTEOLOGICAL TERMS.

Acetabulum (L. a vinegar cruet).—The socket of the
   innominate bone which receives the head of
   the femur.

Acrodont (Gr. *akron*, high ; *odontos*, a tooth).—Having
   the teeth anchylosed to the top of the alveolar
   process.

Acromion (Gr. *akron*, and *omos*, the shoulder).—A process of the scapula forming the summit of
   the shoulder.

Alæ (L. *ala*, a wing.)—Applied to the wing-like processes of bone.

Alinasal process (L. *ala*, and *nasus*, the nose).—A process surrounding the nasal aperture in Amphibians.

Alisphenoid (L. *ala*, and *os sphenoids*).—A distinct bone in certain skulls ; in human anatomy it is represented by the great wings and external pterygoid plates of the sphenoid.

Alveolus (L. *alveolus*, a little hollow).—A hollow or depression, *e.g.* the socket of a tooth.

Amphiarthrosis (Gr. *amphi*, both ; *arthron* a joint).—A class of mixed articulation with partial mobility (see *symphysis*).

Amphicœlus (Gr. *amphi*, both ; *koilos*, hollow).—Applied to vertebræ having a concavity at either end.

Ampulla (L. *ampullor*, I swell out). — The dilated portions of the semicircular canals of the internal ear.

Anapophysis (Gr. *ana*, upon ; *apophuo*, I grow).—A process on the neural arch of certain lumbar vertebræ ; in human anatomy termed accessory process.

Anchylosis(Gr. *ankule*, a thong or clasp).—The union of two or more bones.

Angular (L. *angulus*, a corner).—An element of the mandible in certain Vertebrata.

Angulo-splenial (L. *angulus*, and *splenium*, a splint). An element of the mandible in some Verte-

brata, in human anatomy the name splenial
is given to an ossification taking place in the
membrane on the inner side of Meckel's
cartilage.

Ankylose (see *Anchylosis*).

Annularis (L. *annulus*, a ring).—The fourth digit of
the manus.

Antebrachium (L. *ante*, before ; *brachium*, the arm).
—The fore-arm, composed of radius and
ulna.

Antitrochanter (Gr. *anti*, opposite, against).—In
birds applied to the articular surface of the
ilium upon which the great trochanter of the
femur plays.

Antrum (L. a cavern).—A term applied to deep
cavities in the interior of certain bones.

Apophysis (Gr. *apophuo*, I grow from.)—A process or
protuberance on the surface of a bone.

Appendicular (L. *appendix*, an appendage).—Used to
denote that portion of the skeleton (the
limbs) which is attached to the axial
skeleton.

Aquæductus cochleæ (aqueduct of the cochlea).—A
small canal passing from the cochlea of the
human ear opening inferiorly, close to the
jugular fossa on the under surface of the
petrous bone.

Aquæductus Fallopii (L. aqueduct of Fallopius).—
The canal in the temporal bone transmitting
the facial nerve.

Aquæductus vestibuli (L. aqueduct of the vestibule). —A canal which leads from the vestibule of the internal ear to the posterior surface of the petrous bone.

Articulare (L. *articulus*, a joint).—A bone of the posterior portion of the lower jaw, formed in most vertebrates, except Mammals, from an ossification of Meckel's cartilage.

Arytenoid (Gr. *arytaina*, a pitcher ; *eidos*, shape).— Applied to two cartilages, situated on the upper border of the cricoid cartilage, at the back of the larynx.

Astragalus (Gr. *astragalos*, die-shaped).—A bone of the tarsus, in man forming the ankle-bone, the morphological components consist of tibiale and intermedium.

Atlas (Gr. In the Greek Mythology, a giant who bore up the earth upon his shoulders).—The first cervical vertebra.

Autogenous (Gr. *autos*, self ; *genesis*, birth).—Applied to parts of a bone developed from independent centres of ossification.

Antostylic.—A term applied to skulls in which the mandibular arch is attached to the cranium without the intervention of the hyoid arch.

Axial. A term used to denote that portion of the skeleton forming the main axis of the body.

Axis (L. a pivot).—The second cervical vertebra.

Basalia Gr. *basis*, a pedestal).—Applied to the base of the skull. The basal cartilages of the fins of Elasmobranch fishes.

Basihyal (L. *basis*, the base ; *hyoides*, hyoid bone).—
An ossification of the hyoidean arch in
certain fishes ; represented in man by the
body of the hyoid bone.

Basioccipital (L. *basis*, the base ; *occiput*, the back of
the head).—A bone of the skull ; represented
in man by the basilar process of the occipital
bone.

Basipterygoid (Gr. *basis*, a pedestal ; *pterygion*, a
wing).—A bone of the skull ; represented in
man by the pterygoid plates.

Basisphenoid (Gr. *basis*, a pedestal ; *sphen*, a wedge).—
A bone of the skull ; represented in man by
the posterior portion of the body of the
sphenoid bone.

Basitemporal (Gr. *basis*, a pedestal ; L. *tempora*, the
temples).—One of the bones of the skull.

Basis cranii.—The base or floor of the skull.

Bicuspid (L. *bis*, twice ; *cuspis*, a point).—Having
two points or fangs, as the bicuspid teeth.

Brachium (L. the fore-arm).—The upper arm, that
portion which articulates with the scapula
and extends to the elbow.

Bulla (L. *bulla*, a bubble).—The osseus wall surrounding the tympanum in some Vertebrata.

Calcaneum (L. *calx*, the heel).—The os calcis of the
tarsus, in man forming the heel. It corresponds to the fibulare and perhaps the fibular
sesamoid bone of the morphologist.

Calcar (L. a spur).—The "spur" of some birds.

Canaliculi (L. little canals).—Applied to the minute canals in bone.

Cancellous (L. *cancelli*, trellis-work).—A term applied to certain bone tissue which assumes a spongy form.

Canine (L. *canis*, a dog).—Applied to those teeth next to the incisors, popularly known as "eye-teeth."

Capitulum (L. a little head).—A term applied to a rounded prominence or small head of a bone.

Carina (L. a keel).—A process of the sternum of most birds.

Carpus (Gr. *karpos*, the wrist).—The wrist. That portion of the fore-limb uniting the manus to the fore-arm.

Cartilage (L. *cartilago*, gristle).—One of the animal connective tissues.

Caudal (L. *cauda*, a tail).—Like or relating to the tail, as the caudal vertebræ.

Centrale (L. *centrum*, the centre).—The central bone of the carpus.

Centrum (L. *centrum*, the centre).—The body of a vertebra.

Cerato-hyal (Gr. *keras*, a horn).—An ossification in the lower portion of the hyoidean arch in certain fishes, represented in man by the lesser cornua of the hyoid bone.

Cervical (L. *cervix*, the neck).—Like or relating to the neck, as the cervical vertebræ.

Chevron bones (Fr. a rafter).—Downward processes

of the caudal vertebræ enclosing the posterior portion of the aorta, present in certain Vertebrata.

Chondro-cranium (Gr. *chondros*, gristle ; *kranion*, skull).—The cartilaginous skull of certain lower vertebrates, the embryonic skull of higher vertebrates.

Clavicle ⎫ (L. *clavicula*, a small key).—The collar-
Clavicular ⎭ bone. So-called from its supposed resemblance to an ancient key.

Clinoid processes (Gr. *kline*, a bed ; *eidos*, shape).—Processes of the body of the sphenoid bone.

Cnemial crest (Gr. *kneme*, leg).—A prominent ridge of bone on the proximal end of the tibia.

Coaptation (L. *con*, together ; *apto*, I fit).—Applied to the movement of joints where the articular surface of one bone travels over that of another, as the patella on the femur.

Coccygeal (L. *coccyx*, the cuckoo).—Relating to the coccyx.

Coccyx.—A term applied to the four terminal vertebræ of man, which unite and form a structure which somewhat resembles a cuckoo's beak.

Cochlea (L. a snail's shell).—The interior division of the internal ear.

Condyle (Gr. *kondulos*, a knuckle).—The articular surface of a bone.

Condyles.—The articular surfaces of the ex-occipital bones which articulate with the vertebral column.

Conoid tubercle.—A process on the posterior border of the clavicle in man.

Coracoid (Gr. *korax*, a crow ; *cidos*, shape).—One of the bones of the pectoral girdle in most vertebrates ; represented in man by the coracoid process of the scapula.

Cornicula (L. *corniculum*, a little horn).—A term sometimes applied in human anatomy to the cerato-hyals.

Coronoid (Gr. *korone*, a crow ; *cidos*, shape).—Crow-shaped ; applied to a process of the lower jaw (*coronoid process*).

Costal (L. *costa*, a rib).—Connected with or relating to the ribs, as the costal cartilages.

Cranium (Gr. *kranion*, the skull).—The skull.

Cribriform (L. *cribrum*, a sieve ; *forma*, shape).— Perforated like a sieve. Applied to the perforated portion of the ethmoid bone through which the fibres of the olfactory nerve pass.

Cricoid (Gr. *krikos*, a ring ; *cidos*, shape).—Ring-like. Applied to a series of cartilages surrounding the larynx.

Crista galli (L. *crista*, a crest ; *galli*, of a cock).— Applied to a ridge of the ethmoid bone.

Crista ilii (L. *crista*, a crest ; *ilii*, the ilium).—Applied to one of the borders of the iliac bone.

Crus (L. the leg).—Applied to parts of the body resembling a leg. That portion of the lower extremity lying between the femur and tarsus.

Cuboides (Gr. *kubos*, a cube ; *eidos*, shape).—A bone of the tarsus.

Cuneiforme (L. *cuneus*, a wedge ; *forma*, form).—A bone of the carpus, sometimes termed ulnare; also applied to three bones of the tarsus.

Dentary (L. *dens, dentis*, a tooth).—Relating to the teeth. A bone of the lower jaw containing teeth. In the embryonic human mandible this name is given to an ossification taking place in the membrane on the outer side of Meckel's cartilage.

Dentate (L. *dens, dentis*, a tooth).—Toothed : having short triangular divisions at the margin. Applied to the second vertebra, because of a toothlike process which occurs on it.

Dentine (L. *dens, dentis*, a tooth).—The calcified substance forming part of the tooth, closely allied to bone.

Diaphysis (Gr. *dia*, between ; *phusis*, growth).—The centre of ossification in the shaft of a long bone.

Diapophysis (Gr. *dia*, between ; *apophuo*, to sprout).— The superior transverse process of a vertebra where there are two such processes present.

Diarthrosis (Gr. *dia*, between ; *arthron*, a joint).—A name given to a class of joints possessing considerable yet varying degrees of mobility.

Diastema (Gr. *dia*, apart ; *histemi*, to place).—A space or gap, especially between teeth.

Didactyle (Gr. *dis*, twice ; *daktulos*, a finger).— Possessing two digits.

Digit (L. *digitus*, a finger or toe).—A finger or toe.

Diphycercal (Gr. *diphues*, mixed ; *kerkos*, a tail).—Applied to the tails of fishes when the caudal fin-rays are divided into two equal or nearly equal parts.

Diploe (Gr. *diploös*, double,.—The spongy texture in tabular bones.

Dorsum sellæ.—Applied to the posterior boundary of the *sella turcica.*

Ectopterygoid (Gr. *ektos*, without ; *pteryx*, a wing : *cidos*, shape).—A bone of the skull in some Vertebrata.

Endoskeleton (Gr. *endon*, within).—The bony and cartilaginous portion of the body covered by muscles and integuments.

Ensiform process (L. *ensis*, a sword ; *forma*, form).—Applied to a process of the sternum, also termed *metasternum* and *xiphisternum.*

Entopterygoid (Gr. *entos*, within ; *pteryx*, a wing ; *cidos*, shape).—A bone of the skull in certain fishes (Teleostei).

Epicondyle (Gr. *epi*, upon ; *kondulos*, a knuckle).—The most prominent portion of the internal condyle of the humerus.

Epicoracoid (Gr. *epi*, upon ; *korax*, a crow).—A bone of the pectoral arch in some Vertebrata.

Epihyal (Gr. *epi*, upon ; L. *hyoides*, hyoid bone).—An ossification of the cornua of the hyoidean arch in certain fishes ; represented in man by the stylo-hyoid ligaments.

Epiotic (Gr. *epi*, upon ; *ous, otos*, the ear).—A bone of the skull above the ear. In human anatomy the name is given to one of the osseous deposits in the embryonic petro-mastoid.

Epiphysis (Gr. *epi*, upon ; *phusis*, growth).—Portion of a bone ossified from a separate centre of ossification.

Epipteric (Gr. *epi*, upon ; *pteryx*, a wing).—Applied to a scale-like ossification sometimes found lying between the antero-inferior angle of the parietal bone, and the great wing of the sphenoid bone. Synonymous with a Wormian bone.

Episternum (Gr. *epi*, upon ; *sternon*, the breast).—A median membrane bone, connected in some Vertebrata with the sternum; also termed the *interclavicle*.

Ethmoid (Gr. *ethmos*, a sieve ; *eidos*, shape).—A bone of the skull having its posterior surface perforated, forming the cribriform plate.

Ethmo-turbinals (Gr. *ethmos*, a sieve ; L. *turbo*, a twining round).—Applied to certain folded bones entering into the formation of the olfactory chambers of the skull.

Ethmovomerine plate.—The anterior plate formed by the union of the trabeculæ in the fœtal skull.

Ex-occipital (L. *ex*, without ; *occiput*, the head).—A bone of the skull, lying on either side of the foramen magnum ; in man formed by the condyloid portion of the occipital bone.

Femoral (L. *femur*, the thigh).—Relating to the femur.

Femur (L.)—The thigh bone.

Fenestra ovalis (L. oval window).—An oval shaped aperture in the tympanum of the ear into which the stapes (or its homologue in the lower Vertebrata) fits.

Fenestra rotunda (L. round window).—A small round aperture in the tympanum of the ear.

Fibula (L. *fibula*, a buckle).—The outer and smaller bone of the leg.

Fibulare.—A bone of the tarsus articulating with the fibula, sometimes termed *calcaneum*.

Fontanelle (L. *fons*, a fountain).—A membranous interval between the bones of the skull, through which arterial pulsation may be seen, hence its name.

Foramen (L. an aperture or opening).—A perforation or opening by which blood vessels or nerves pass through the bones.

Foramina, *plu.*

Foramina incisiva (L. incisor openings).—Openings that remain in the Mammalia between the premaxillæ and palatine plates of the maxillary bones.

Foramina obturatoria (L. openings to be stopped up).—The apertures in the innominate bones between the ischia and pubes, and which are covered by fibrous membrane.

Fossa (L. *fossa*, a ditch).—A small cavity or depression in a bone.

Frontal (L. *frons, frontis*, the forehead).—A bone of the skull relating to the region of the forehead.

Furculum (L. *furca*, a fork).—A bone in birds formed by the united clavicles, and which is somewhat V-shaped.

Ginglyform (Gr. *ginglumos*, a hinge).—Hinge-like.

Ginglymus.—Applied to joints which allow motion in two directions only, as the joint of the elbow.

Glabella (L. *glabra*, smooth).—An eminence between, and connecting together, the two superciliary ridges ; sometimes termed nasal eminence.

Glenoid fossa (Gr. *glene*, a socket ; *eidos*, shape).—The cavity in the scapula in which the head of the humerus articulates.

Gomphosis.—The articulation of a tooth with its socket.

Hæmal arch (Gr. *haima*, blood).—Applied to the arch under the vertebral column, which encloses and protects blood vessels.

Hæmapophyses (Gr. *haima*, blood ; *apophuo*, I grow from).—Processes of the vertebræ which form the hæmal arch.

Hallux (L. *hallux*, the great toe).—The first digit of the pes.

Hamulus lachrymalis (L. *hamulus*, a little hook).—The hook-like process of the lachrymal bone.

Harmonia (Gr. *harmoza*, I fit together).—A variety of suture in which the two bones are in simple apposition.

Hiatus Fallopii.—A groove on the internal surface of the petrous portion of the temporal bone, which lodges the great superior petrous nerve.

Homocercal (Gr. *homoios*, like; *kerkos*, a tail).—Applied to the tails of fishes when the caudal fin-rays are arranged symmetrically to the axis of the body.

Humerus (L. the shoulder).—The bone of the arm from the shoulder to the elbow.

Hyoid (Gr. *v* (the letter upsilon); *eidos*, shape).—A small bone situated between the tongue and the larynx.

Hyomandibular (Gr. *hyoides*, hyoid bone; L. *mandible*, the lower jaw).—The cartilage or bone at the proximal end of the hyoidean arch.

Hyoplastron (Gr. *plastos*, moulded).—In the Chelonia the second lateral piece of the plastron.

Hyostylic.—A term applied to skulls in which the mandibular arch is supported by the hyoid arch.

Hypapophysis (Gr. *hupo*, under; *apophuo*, I grow from).—A process on the ventral surface of the vertebræ in some Vertebrata.

Hypoplaston (Gr. *hupo*, under).—In the Chelonia the third lateral piece of the plastron.

Hypotarsus (Gr. *hupo*, under; *tarsos*, the flat of the foot).—A process developed in most birds from the region of the tarsal and metatarsal bones.

Hypural (Gr. *hupo*, under ; *oura*, the tail).—The bones which in fishes support the caudal fin-rays.

Iliac (L. *ilia*, the flanks).—Relating to the region of the ilium.

Ilium (L. *ilia*, the flanks).—A bone of the pelvic girdle ; in the higher Vertebrata the ilium forms the upper division of the os innominatum.

Incisor (L. *incido*, I cut into).—A cutting tooth.

Incisura of acetabulum (L. *incisura*, a notch).—An incomplete portion of the margin of the acetabulum.

Incisura ethmoidalis. — A notch separating the orbital plates of the frontal bone.

Incisura semilunaris.—The centre or middle notch in the upper border of the sternum.

Incus (L. an anvil).—A small ossicle of the ear.

Index (L. *indico*, I point out).—The index or forefinger, the second digit of the manus.

Infra-orbital (L. *infra*, beneath).—One or more bones beneath the orbit in fishes.

Infra-temporal crest.—A ridge bounding the temporal fossa inferiorly and separating it from the pterygoid fossa. Synonymous with pterygoid ridge.

Innominate (L. *in*, not ; *nomen*, a name).—The bone forming the pelvis of higher Vertebrata.

Interclavicula L. *inter*, between ; *clavicula*, the collar-bone).—A median membrane bone between the clavicles in many Vertebrata, also termed the *episternum*.

Intercrural (L. *inter*, between).—A term applied to additional neural arches, when more than one is present on each vertebra.

Intermaxillary (L. *inter*, between).—A name sometimes applied to the premaxillary bones.

Intermedium (L. *intermedius*, intermediate).—A bone of the carpus, also termed *centrale*.

Interoperculum (L. *inter*, between ; *operculum*, a lid).—A bone in fishes lying between the operculum and preoperculum.

Interspinous (L. *inter*, between).—Between the spines. Applied particularly to certain bones on the dorsal fin of fishes developed between the spines of the vertebræ.

Ischium (Gr. *ischion*, the hip).—A bone of the pelvic girdle.

Jugal (L. *jugum*, a yoke).—A name applied to the malar or cheek bone.

Jugular (L. *jugulum*, the fore-part of the neck).—Pertaining to the neck. Used to denote the position of ventral fins in some fishes, when they are situated in front of the pectoral fins.

Lachrymal (L. *lachryma*, a tear).—Applied to a bone on either side of the face, through which the lachrymal duct passes from the eye to the nostrils ; the *os unguis* of some authors.

H

Lambdoidal suture (Gr. the letter Λ, *lambda*; *eidos*, shape).—Another name for the occipito-parietal suture of the skull.

Lamina spiralis (L. spiral plate).—The thin, bony septum of the cochlea of the ear.

Limbous (L. *limbus*, a border).—A term applied to such sutures as that between the parietal and frontal bones.

Linea aspera (L. rough line).—A prominent ridge extending along the femur.

Lingula sphenoidalis (L. *lingua*, a tongue).—A small bony spicule of the sphenoid bone bounding the carotid groove externally, it has a separate centre of ossification and corresponds to the sphenotic of lower Vertebrata.

Lophosteon (Gr. *lophos*, neck; *osteon*, a bone).—That portion of the sternum in birds from which the keel ossifies.

Lunare (L. a little moon).—A bone of the carpus, also termed the *semilunare* or *intermedium*.

Malar (L. *malar*, the cheek).—The cheek bone, also termed *jugal*.

Malleolar (L. *malleolus*, a little hammer).—A bone in the Ruminantia, articulating below with the calcaneum, and above with the astragalus.

Malleolus (L. a little hammer).—A process of the tibia and fibula.

Malleus (L. a hammer).—One of the three movable bones of the ear, articulating with the tympanic membrane on the one side and the incus on the other.

Mandible (L. *mando*, I chew).—The lower jaw, the inferior maxilla of human anatomy.

Manubrium (L. a handle).—Applied to the anterior portion of the sternum, also termed *presternum*.

Manus (L. the hand).—The hand.

Marginal bones.—A name given to certain additional bones on the radial and ulnar borders of the manus in the *Ichthyosauria*.

Mastoid (Gr. *mastos*, a nipple ; *eidos*, shape).—Applied to a process of the temporal bone behind the ear.

Maxilla (L. a jaw).—The upper jaw, although sometimes applied to both upper and lower jaws, which are then spoken of as inferior and superior maxillæ respectively.

Maxillo-turbinal.—The inferior turbinated bone of the face.

Meatus (L. a way, passage).—A small canal.

Meatus auditorius externus (L. external auditory canal). — The passage leading from the exterior to the cavity of the tympanum.

Meatus auditorius internus (L. internal auditory canal).—The opening in the posterior portion of the periotic, for the passage of the auditory nerve to the internal ear.

Meckel's cartilage.—The cartilage of the chondocranium forming the axis of the mandible.

Medius (L. middle).—The third digit.

Mental prominence (L. *mentum*, the chin).—A triangular elevation of the mandible forming the chin, and peculiar to the human skull.

Mental spines.—A name applied to two pairs of prominent tubercles on the mandible for the attachment of muscles.

Mento-meckelian element (L. *mentum*, the chin).—Applied to that portion of Meckel's cartilage which becomes ossified and forms the chin.

Mesethmoid (Gr. *mesos*, middle; *ethmos*, a sieve; *eidos*, shape).—The vertical plate of the ethmoid bone in man, a cartilage in the lower Vertebrata.

Mesopterygium (Gr. *mesos*, middle; *pterux*, a wing).—The median basal cartilage in the pectoral fin of the Elasmobranchii.

Meso-pterygoid (Gr. *mesos*, middle; *pterux*, a wing).—A bone of the skull in certain fishes (Teleostei).

Meso-scapula (Gr. *mesos*, middle; L. *scapula*, the shoulder-blade).—A term sometimes applied to the spine of the scapula.

Meso-sternum (Gr. *mesos*, middle; *sternon*, the chest).—The middle portion of the sternum, sometimes termed the *gladiolus*.

Metacarpal (Gr. *meta*, beyond; *karpos*, the wrist).—Applied to each bone of the metacarpus.

Metacarpus (Gr. *meta*, beyond; *karpos*, the wrist).—That part of the manus situated between the wrist and the fingers.

Metacromion (Gr. *meta*, beyond; *akron*, the summit;

omos, a shoulder).— The long projecting process from the acromion process found in certain rodents.

Metapophysis (Gr. *meta*, beyond ; *apophuo*, I grow from).—The mammillary process of human anatomy. A lateral process sometimes developed on the vertebræ.

Metapterygium (Gr. *meta*, behind ; *pterux*, a wing).— The hinder basal cartilage in the pectoral fin of the Elasmobranchii.

Metasternum (Gr. *meta*, behind ; *sternon*, the chest).— The *xiphisternum*. The posterior portion of the sternum, also termed the ensiform cartilage.

Metatarsal (Gr. *meta*, beyond ; *tarsos*, the sole of the foot).—Applied to each bone of the metatarsus.

Metatarsus (Gr. *meta*, beyond ; *tarsos*, the sole of the foot).—That part of the pes situated between the ankle and the toes.

Metosteon (Gr. *meta*, behind ; *osteon*, a bone).—The posterior portion of the sternum in birds.

Minimus (L. the least).—The fifth digit of the *pes* or *manus*.

Molar (L. *mola*, a mill).—A grinding tooth.

Multicuspid (L. *multis*, many ; *cuspis*, a pointed extremity).—Possessing many cusps, as the molar teeth.

Mylo-hyoid groove.—A groove on the internal side of the mandible, in which the mylo-hyoid nerve and artery run.

Murtiform fossa.—A depression on the superior maxillary bone, known also as the *incisor fossa*.

Nasal (L. *nasus*, the nose).—Relating to the nose.

Naviculare or Navicular (L. *navicula*, a little ship).—A bone of the tarsus, also termed *centrale*, and also used for the *scaphoid* of the carpus.

Necrosis (Gr. *neckros*, dead).—The death of a mass of bone.

Neural plate.—Cartilaginous plates forming the sides of the neural canal in the Elasmobranchii.

Neuropophysis (Gr. *neuron*, a nerve; *apophuo*, I grow from).—The spinous process of a vertebra formed at the point of the neural arches.

Neuro-central suture (Gr. *neuron*, a nerve; L. *sutura*, a seam).—The suture between the ossification of the centrum of a vertebra and those of the neural arches.

Obturator foramen.—The aperture between the ischium and pubis, closed by fibrous membrane.

Occipital (L. *occiput*, the back part of the head).—Pertaining to, or connected with, the back part of the head.

Occipito-mastoid suture.—A continuation of the lambdoidal suture.

Occipito-parietal suture.—The suture between the occipital and parietal bones, also termed lambdoidal.

Occiput (L. *occiput*, the back part of the head).—The hinder part of the head or skull.

Odontoid (Gr. *odous*, a tooth ; *eidos*, shape).—Applied to a process, tooth-like in shape, of the second cervical vertebra.

Olecranon (Gr. *olene*, the elbow ; *kranion*, the top of the head).— The summit of the ulna, which forms the elbow.

Olivary eminence (L. olive-shaped).—An eminence in mammals on the superior surface of the body of the basi-sphenoid.

Omosternum (Gr. *omos*, the shoulder ; *sternon*, the breast).—A median process developed from the coraco-scapula cartilages in the Batrachia.

Operculum (L. a lid).—The gill cover or bony flap covering and protecting the gills in many fishes.

Opisthocœlus (Gr. *opisthen*, behind ; *koilos*, hollow).— Applied to those vertebræ the centra of which are concave posteriorly.

Opisthotic (Gr. *opisthen*, behind ; *ous, otos*, the ear).— An ossification of the temporal bone surrounding the fenestra rotunda and cochlea. In Teleostean fishes it persists as a separate bone.

Orbito-sphenoid (L. *orbito*, the orbit ; Gr. *sphen*, a wedge ; *eidos*, shape).—A bone of the skull in front of the optic foramen, in human anatomy termed the lesser wing of the sphenoid bone.

Os articulare (L. articular bone).—A bone in the mandible in some Vertebrata.

Os calcis (L. the heel-bone).—Another name for the *calcaneum.*

Os cloacæ (L. the bone of the cloaca).—A bone supporting the front wall of the cloaca in some Lacertilia.

Os coxæ (L. the hip bone).—Another name for the *innominate* bone of the pelvic girdle.

Os en ceinture (Fr. girdle bone).—A peculiar cartilage bone in the frog's skull.

Os innominatum (L. a nameless bone).—In the higher Vertebrata the large bone of the pelvis, formed by the coalescence of the ilium, ischium, and pubis.

Os linguæ (L. bone of the tongue).—Synonymous, with the hyoid bone.

Os magnum (L. large bone).—A bone of the carpus known also as the *capitatum.*

Os odontoideum (L. tooth-shaped bone).—The peg-like bone, anchylosed with the second verte-bra in the higher Vertebrata, on which the head rotates.

Os orbiculare or lenticulare (L. orbicular or lentil shaped bone).—A name formerly used for the tubercle of the incus, in human anatomy known as the lenticular process of the incus.

Os penis (L. penis bone),—A bone in the penis of certain Vertebrata.

Os planum (L. flat bone).—Another name for the orbital surface of the ethmoid bone, or the external boundary of the ethmo-turbinals.

Os pubis (L. pubic bone).—A bone of the pelvis. In the higher Vertebrata a portion of the *os innominatum*.

Os trigonum (Gr. triangular bone).—The posterior part of the astragalus. Frequently developed from a separate centre of ossification.

Ossa suprasternalia (L. bones of the sternum).—Applied to two small nodules of bone sometimes found in connection with the presternum; they are supposed to represent median portions of the episternum.

Ossa triquetra (L. triangular bones).—Small triangular pieces of bone sometimes present between the bones of the skull.

Ossa wormii (after Wormian, a Danish anatomist).—Bones found in sutures such as the *ossa triquetra*, sometimes termed *ossa suturarum*.

Osseus (L. *os*, a bone).—Composed of, or resembling bone.

Ossicules.—Small bones.

Ossicula auditus (L. little hearing bones).—The three small bones in the tympanum of the ear.

Ossification (L. *os*, a bone ; *facio*, I make).—The process by which cartilage or membrane is converted into bone.

Osteoblast (Gr. *osteon*, a bone ; *blastos*, a germ).—The bone-cells which probably secrete the *osteogen*.

Osteoclast (Gr. *osteon*, a bone ; *klao*, I break).—A term applied to the large nucleated cells which form excavations in bony tissue during the process of absorption.

Osteodentine (Gr. *osteon*, a bone ; L. *dentis*, a tooth).
—A substance intermediate in structure
between dentine and bone.

Osteogen (Gr. *osteon*, a bone ; *gennao*, I produce).—
The soft transparent tissue which in the
formation of bone becomes ossified.

Osteology (Gr. *osteon*, a bone ; *logos*, a discourse).—
That branch of anatomy which treats of the
skeleton of vertebrated animals.

Palatine (L. *palatum*, the palate).—The name of a
pair of bones in the skull, also of *foramina* in
the same bones.

Parachordal (Gr. *para*, by, near ; *chorde*, a string).—
A term applied to the embryonic cartilages
in the posterior portions of the skull.

Parapophysis (Gr. *para*, beside ; *apophuo*, I grow
from).—The inferior transverse process of a
vertebra, when two are present on each side
of the centrum ; also called the *capitular
process* from the fact that the head of the rib
articulates here.

Parasphenoid (Gr. *para*, beside ; *sphen*, a wedge ;
eidos, shape).—A bone of the skull in some
Vertebrata, forming portion of the base
between the basi-occipital and pre-sphenoidal
region.

Parietal (L. *paries*, a wall). The name of a pair of
bones in the skull.

Parieto-mastoid suture.—The suture between the
parietal bone and the mastoid portion of the
temporal.

Paroccipital (Gr. *para*, beside ; *occipito*, the head).—
A name given to a pair of processes of the
ex-occipital bones, sometimes termed *para-
mastoid*, and in man *jugular* processes.

Patella (L. *patella*, a small pan or plate).—The knee-
cap or pan, also termed the *rotula*.

Pectoral (L. *pectus*, a breast).—Connected with or
belonging to the region of the chest, *e.g.*, the
pectoral arch.

Pedicle (L. *pediculus*, a little foot).—The ventral
portion of either side of the neural arch of a
vertebra.

Pelvis (L. *pelvis*, a basin).—The cavity surrounded by
the pelvic girdle.

Perichondrium (Gr. *peri*, around ; *chondros*, gristle).—
The sheath of connective tissue covering
cartilages.

Pericranium (Gr. *peri*, around ; *kranion*, the skull).—
The region around the skull.

Periosteum (Gr. *peri*, around ; *osteon*, a bone).—The
sheath of connective tissue which invests
bones.

Periotic (Gr. *peri*, around ; *ous, otos*, the ear).—A
name given to the petro-mastoid portion of
the temporal bone.

Pes (L. foot).—That portion of the hind limb which
includes the tarsus, metatarsus, and digits, in
man forming the foot.

Petrous (Gr. *petros*, a stone).—Applied to a dense,
solid mass of bone forming the basal part of
the temporal bone.

Phalanges (Gr. *phalanx*, a line of soldiers).—A term applied to the bones of the digits, they being arranged in rows. (Singular *phalanx*.)

Pisiform (L. *pisum*, a pea ; *forma*, shape).—A name applied to the small bone of the Mammalia, situated in the tendon of the flexor muscle of the wrist, on the ulna side. It is supposed by some anatomists to be the remains of a seventh digit.

Planiform (L. *planum*, a level place ; *forma*, shape).— Applied to a class of joints in which the surfaces are nearly flat ; also termed arthrodial.

Plastron (Gr. *emplastron*, a plaster).—The ventral exoskeleton of the Chelonia.

Pleurapophysis (Gr. *pleuron*, a rib ; *apophuo*, I grow from).—A lateral out-growth from a vertebra.

Pleurosteon (Gr. *pleuron*, a side ; *osteon*, a bone).—A lateral ossification in the anterior portion of the sternum of birds.

Pollex (L. thumb).—The first digit of the manus, the thumb.

Post-axial.—A term applied to that surface of a limb, which is posterior, when the limb is at right angles to the vertebral column.

Post-clavicula (L. *post*, after ; *clavicula*, the collar bone).— A bone of the pectoral arch in certain Teleostean fishes.

Posterior (L. following after).— Towards the tail ; behind.

Post-frontal (L. *post*, after).—A bone in some Vertebrata which lies behind the orbit and dorsal
to the ali-sphenoid.

Post-glenoid process.—A process of the temporal
bone, situated immediately behind the
glenoid fossa.

Post-orbital process.—A process on the dorso-lateral
border of the frontals in some Vertebrata
for the attachment of muscles.

Post-sphenoid (L. *post*, after).—An independent ossification of the posterior portion of the body
of the sphenoid bone, seen in the fœtal
skull.

Pre-axial (L. *præ*, before).—A term applied to that
surface of a limb, which is anterior, when
the limb is at right angles to the vertebral
column.

Pre-coracoid (L. *præ*, before).—A bone of the pectoral arch in some Vertebrata.

Pre-frontal (L. *præ*, before).—A bone of the skull in
certain Vertebrata.

Pre-maxilla (L. *præ*, before).— A bone formed in
most skulls in which the upper incisor teeth
are implanted. It is present in the human
fœtus, but fuses normally with the maxilla
before birth.

Pre-operculum (L. *præ*, before).—A bone in Teleostean fishes, situated in front of the operculum.

Pre-sphenoid (L. *præ*, before).— A bone of the skull
in some Vertebrata in front of the sphenoid

bone. In human osteology this name is given to the anterior portion of the body of the developing sphenoid bone.

Pre-sternum.—The *manubrium*, which see.

Process (L. *processus*, to go forwards).—A projecting part or protuberance of a bone.

Processus brevis vel obtusus (L. short or obtuse process).—Applied to a short process on the malleus.

Processus cochleariformis (L. spoon-shaped process).—A thin plate of bone in the petrous portion of the temporal bone, situated above the Eustachian canal.

Processus gracilis (L. slender process).—Applied to the long process on the malleus.

Processus lenticularis (L. lentil-shaped process).—The *os orbiculare*, a process of the incus which articulates with the stapes.

Procœlus (Gr. *pro*, before ; *koilos*, hollow).—Applied to vertebræ which have a concavity in their centra in front.

Prognathous (Gr. *pro*, forward ; *gnathos*, a jaw).—Applied to skulls with a large cranio-facial angle and protruding upper jaw.

Pro-otic (Gr. *pro*, before ; *ous, otos*, the ear).—A bone in some Vertebrata in front of the ear. In human anatomy the name is given to one of the osseous deposits in the embryonic petromastoid.

Propterygial (Gr. *pro*, before ; *pterux*, a wing).—A

name applied to the anterior basal fin cartilages in the Elasmobranchii.

Pterotic (Gr. *pteron*, a wing ; *ous, otos*, the ear).—An ossification between the pro-otic and epi-otic bones in some Vertebrata. The name is given by some human anatomists to one of the osseous deposits in the embryonic petromastoid.

Pterygoid (Gr. *pteron*, a wing ; *eidos*, shape).—A pair of bones behind or lying partly on the palatines of some Vertebrata ; in human anatomy they are termed the internal pterygoid plates of the sphenoid bone.

Pubis (L. *pubes*, the region of the groin).—A bone of the pelvis ; in human anatomy the anterior portion of the *os innominatum*.

Pygostyle (Gr. *puge*, the rump ; *stulos*, a style, pen).— A bone in most birds supporting the tail feathers ; the ploughshare bone.

Quadrate (L. *quadratus*, square).—A bone of the skull articulating with the lower jaw in all Vertebrata below the Mammalia.

Quadrato-jugal (L. *quadratus*, square ; *jugum*, a yoke).—A bone formed by the union of the quadrate and jugal in some Vertebrata, in others a bone situated behind the jugal and maxillary bones.

Quadrato-palatine (L. *quadratus*, square ; *palatus*, the pallet).—A bone formed by the union of the quadrate and palatine in some Vertebrata.

Radial sesamoid. — The tubercle of the human scaphoid. Supposed by some to represent the first digit of the heptadactylus manus, *i.e.*, a pre-pollex.

Radius (L. *radius*, a ray, or spike). — The outer bone of the fore-arm.

Ramus (L. *ramus*, a branch). — Each half of the lower jaw or mandible of the Vertebrata.

Rostrum (L. *rostrum*, a bill or beak). — Applied to the anterior or facial portion of the skull of certain fishes, where greatly produced. A sharp prominence or spine in the middle line of the sphenoid bone of the skull.

Sacral (L. *sacrum*, sacred). — Relating to the region of the sacrum.

Sacrum. — The terminal anchylosed vertebræ, immediately succeeding the lumbar vertebræ, to which the pelvic arch is articulated.

Sagittal suture (L. *sagitta*, an arrow ; *sutura*, a seam). — The suture uniting the parietal bones of the skull.

Scaphoid (Gr. *scaphe*, a boat : *eidos*, shape). — Boat-shaped.

Scaphoides (Gr. *scaphe*, a boat ; *eidos*, shape). — One of the bones of the carpus, and also one of the tarsus.

Scapho-lunar (Gr. *scaphe*, a boat ; L. *lunar*, a moon). —

Scapula (L. the shoulder-blade). — The shoulder-blade of the pectoral arch of Vertebrata.

Scapula accessoria (L. additional scapula). — A small

bone on the outer side of the scapula. Developed in some birds.

Schindylesis (Gr. *schindulio*, I split).—A variety of suture where one bone fits into a groove in another.

Sella turcica.—The depression on the superior surface of the basi-sphenoid which lodges the pituitary body.

Sensory canal bones.—Applied to a series of ossifications which protect and convey the sensory canals in certain fishes.

Sesamoid (Gr. *sesamon*, a kind of small grain ; *eidos*, shape).—Applied to small bones formed in tendons.

Sinciput (the fore part of the head).—The fore part of the head or skull.

Sphenethmoid (Gr. *sphen*, a wedge ; *ethmos*, a sieve ; *eidos*, shape).—A bone of the skull in some Vertebrata, anterior to the parasphenoid.

Spheno-palatine foramen.—A foramen formed by the palatine and sphenoid bones.

Sphenoid (Gr. *sphen*, a wedge ; *eidos*, shape).—A bone of the skull which wedges in and locks together other bones.

Spinous (L. *spina*, a thorn).—Applied to any spine-like process of a bone).

Splenial (L. *splenium*, a splint).—A bone of the skull in certain Vertebrata.

Squamosal (L. *squama*, a scale).—A bone of the skull in lower Vertebrata, represented in human

1

anatomy by the squamous part of the temporal bone.

Squamo-zygomatic (L. *squama*, a scale ; Gr. *zugnum*, I yoke together).—A suture which forms a centre of ossification in the fœtal skull.

Stapes (L. a stirrup).—A stirrup-like bone of the ear, fitting in the fenestra ovalis.

Sternum (Gr. *sternon*, the breast).—The flat bone of the breast to which the rib or costal cartilages are attached.

Stylo-hyal.—An ossification in certain Vertebrata, which connects the hyoidean arch with the skull. The styloid process of the temporal bone of human anatomy consists of a fusion of *Stylo-hyal* and *Tympano-hyal*.

Styloid (Gr. *stulos*, a style ; *eidos*, shape).—Style-like. Applied to certain processes of bones, as those of the ulna, radius, and temporal.

Sub-operculum (L. *sub*, under ; *operculum*, a lid).—A bone in Teleostean fishes forming the ventral portion of the skeleton covering the gill.

Suchospondylia (Gr. *suchnos*, long ; *spondulos*, a vertebra). — Applied to certain Reptilia having elongated and divided transverse processes on the dorsal vertebræ.

Sulcus (L. a furrow).—A term applied to furrows or depressions in bones, *e.g.* the *sulcus frontales* of the frontal bone.

Superciliary ridge (L. *super*, above ; *cilium*, an eyelash).—An eminence on the frontal bone, above the margin of the orbit, caused by the projection of the frontal sinus.

Supra-angular (L. *supra*, above).—A bone of the mandible in some Vertebrata, above the angular.

Supraclavicle (L. *supra*, above ; *clavicula*,. the collar-bone).—A bone at the dorsal end of the clavicle. Present in some fishes.

Supracondyloid process (L. *supra*, above).—A small process sometimes present on the humerus. It represents a portion of the bar of bone which in many animals passes from this situation to the *ento-condyle* and forms a *supra-condyloid* foramen, through which the median nerve and the brachial artery pass.

Supra-occipital (L. *supra*, above ; *occiput*, the back of the head).—A bone of the skull above the foramen magnum in some Vertebrata.

Supra-orbital (L. above the orbit).—A bone of the skull in some Vertebrata, also applied to a notch or foramen above the orbit in the frontal bone.

Supra-scapular.—Applied to the superior border of the scapula. In some animals there is a distinct cartilage in this situation.

Supra-temporal.—A bone of the skull in certain Vertebrata.

Suspensorium (L. *suspendo*, I hang, or suspend).—The apparatus suspending the lower jaw to the cranium.

Suture (L. *sutura*, a seam).—An immovable articulation of bones.

Symphysis (Gr. *sun*, together ; *phusis*, I grow).—The union of bones where there is not a complete articulation.

Symplectic (Gr. *symplino*, to entwine together).—A bone of the skull in Teleostean fishes, forming the lower ossification of the suspensorium and articulating below with the quadrate.

Synarthrosis (Gr. *sun*, together ; *arthron*, a joint).—A class name for immovable articulations.

Synchondrosis (Gr. *sun*, together ; *chondros*, gristle).—A form of articulation existing in early life, in which a thin layer of cartilage is interposed between the bones.

Synostosis (Gr. *sun*. together ; *osteon*, a bone). — Applied to the premature obliteration of the sutures of the skull.

Synovia (Gr. *sun*, together ; *oon*, an egg).—A viscid, transparent fluid, secreted in the cavities of joints.

Synovial membrane.—A thin membrane situated in the interior cavity of a joint, covering the inner surfaces of the ligaments connected with the joint. The membrane secretes the synovia.

Talus (L. a die).—The astragalus, which see.

Tarsalia (Gr. *tarsos*, the flat of the foot).—The bones of the tarsus.

Tarso-metatarsus.—The single bone in the leg of a

bird, being formed by an anchylosis of the tarsus and metatarsus.

Tarsus (Gr. *tarsos*, the flat of the foot).—That portion of the posterior limb between the crus and metatarsus, the ankle in man.

Temporal (L. *tempora*, the temples).—The names of each of a pair of bones in the skull of higher Vertebrata ; the squamosal of many lower Vertebrata.

Temporo-parietal.—Applied to the suture joining the temporal and parietal bones of the skull.

Tetradactyle (Gr. *tetras*, four ; *daktulos*, a finger or toe).—Possessing four fingers.

Thecodont (Gr. *theke*, a sheath ; *odous*, *odontos*, a tooth).—Having the teeth lodged in alveoli, as in the Protorosauria, a group of extinct Lacertilia.

Thyro-hyal (G. *thureus*, a shield).—Applied to two ossifications of the hyoid in the lower Vertebrata ; homologous with the great cornua of the hyoid bone in man.

Thyroid (Gr. *thureos*, a shield ; *eidos*, shape).—Applied to the largest cartilage of the larynx.

Tibia (L. a flute).—The larger of the two bones of the leg.

Tibiale.—A bone of the tarsus articulating with the tibia.

Trabecula (L. a little rafter).—Two cartilaginous plates in the embryo from which the dorso-anterior region of the skull arises.

Trapezium (Gr. *trapezion*, a geometrical figure, from *trapeza*, a small table or board).—A bone of the carpus of very irregular form.

Trapezoid (Gr. *trapeza*, a table ; *eidos*, shape).—One of the carpal bones.

Tricuspid (L. *tria*, three ; *cuspis*, a point).—Having three points.

Tridactyle (L. *tria*, three ; Gr. *daktulos*, a finger or toe). —Having three digits.

Tridentate (L. *tria*, three ; *dens*, *dentis*, a tooth).— Having three tooth-like divisions.

Trochanter (Gr. *trochas*, I roll or run round).—One of the two processes (*major* or *minor*) on the upper part of each femur.

Trochlea (L. *trochlea*, a case containing pulleys).— Applied to that part of the humerus articulating with the ulna.

Tuber calcis (L. *tuber*, a knob ; *calx*, the heel).—The large posterior extremity of the *os calcis*.

Tuberosity (L. *tuber*, a knob).—A term applied to any knob-like bony prominence, usually giving attachment to muscles.

Tympanic (L. *tympanum*, a drum).—A bone in some Vertebrata surrounding the tympanum of the ear.

Ulna (Gr. *olene*, the elbow).—The bone of the fore-arm, the proximal end of which forms the elbow in man.

Ulnare.—A bone of the carpus articulating with the ulna.

Unciform (L. *uncus*, a hook ; *formis*, shape).—A bone of the carpus also termed *uncinatum* and *hamatum*.

Uncinate (L. *uncus*, a hook).—Hooked. A process of the ethmoid bone.

Ungual phalanges (L. *unguis*, a nail).—The terminal phalanges of the digits which are provided with nails or claws.

Urohyal (Gr. *oura*, the stern, the tail).—An ossification in the posterior portion of the hyoidean arch in fishes.

Urosacral (Gr. *oura*, the tail ; *sacrus*, sacred).—A term applied to certain caudal vertebræ in birds which are anchylosed together.

Urostyle (Gr. *oura*, tail ; *stulos*, a style, or pen).—A bony prolongation of the vertebral column in certain fishes and amphibia.

Venter of scapula (L. *venter*, the belly).—A broad concavity on the anterior surface of the scapula ; also termed the *sub-scapular fossa*.

Vertebra (L. *verto*, to turn).—The name of each of the bony segments forming the spinal column.

Vertebrata dentata.—A term sometimes applied to the second cervical vertebra or axis.

Vertebra prominens.—A term sometimes applied to the seventh cervical vertebra in man, on account of its prominent spine.

Vidian canal (after *Vidius*, a French Professor).—A small canal which passes through the internal pterygoid plate of the sphenoid bone, trans-

---mitting a branch of the internal maxillary artery and a nerve passing from the spheno-palatine ganglion.

Vomer (L. a ploughshare).—A bone of the skull, so-called from its fancied resemblance to a ploughshare.

Weberian ossicles (after Weber, a German professor).—A series of three ossicles, originally described by Weber, serving as a connection between the internal ear and the air-bladder in certain families of Physostomous Teleostei.

Wormian bones.—Bones found in sutures, sometimes termed *ossa suturarum*.

Xiphisternum (Gr. *xiphos*, a sword; *sternon*, the breast).— The posterior portion of the sternum, in Human Anatomy known as the *ensiform cartilage*.

Xiphoid cartilage (Gr. *xiphos*, a sword; *eidos*, shape). —Sometimes applied to the *xiphisternum*.

Zygapophyses (Gr. *zugos*, a yoke ; *apophuo*, I grow from).— The articulating process of the vertebræ.

Zygomatic arch
Zygoma (Gr. *zugos*, a yoke).—The arch formed by the zygomatic process of the temporal bone, and the malar bone.

Zygomatic fossa (Gr. *zugos*, a yoke ; L. *fossa*, a ditch).—The lower part of the fossa bridged over by the zygomatic arch.

INDEX.

INDEX.

BIRMINGHAM :

PRINTED BY ROBERT BIRBECK & SONS, BROAD STREET.

9 783741 187049